Revolutionary Bio-politics from Fedorov to Mao

Jeff Love · Michael Meng

Revolutionary Bio-politics from Fedorov to Mao

palgrave
macmillan

Jeff Love
Initiative for the Study of Russian
Philosophy and Religious Thought
Northwestern University
Evanston, IL, USA

Michael Meng
Department of History
Clemson University
Clemson, SC, USA

ISBN 978-981-99-4744-7 ISBN 978-981-99-4745-4 (eBook)
https://doi.org/10.1007/978-981-99-4745-4

Cover credit: Maram_shutterstock.com

This Palgrave Macmillan imprint is published by the registered company Springer Nature Singapore Pte Ltd.
The registered company address is: 152 Beach Road, #21-01/04 Gateway East, Singapore 189721, Singapore

Contents

1	**Introduction**	1
	1 Universal Resurrection	3
	2 The Infinite	9
	3 Chapter Outline	14
	4 Our Approach	15
2	**Becoming God**	21
	1 Divine Humanity	25
	2 Universal Resurrection	38
	3 The Power of Fantasy	49
3	**Recuperation of the Finite**	55
	1 The Nothing	56
	2 Transcendence	64
	3 Genuine Thinking	70
	4 Historie and Geschichte	73
	5 Technology and Nihilism	80
4	**Reconciliation—The Great Harmony**	87
	1 Mao Zedong	91
	2 Mou Zongsan	97
	3 Yuk Hui	107

5 Epilogue: Harmony with Suffering? 117
1 Terror 118
2 Alien Intelligence 120
3 Ordinary and Awful 123
4 Clinging to Life 126

Index 129

CHAPTER 1

Introduction

Our world is terrifying and cruel. We wake up to death every day, of loved ones, strangers, children and the elderly, warriors and pacifists, addicts of health regimes and drugs. No matter where we look, we cannot escape, even for a moment, the distressing spectacle of mortal change. We may choose to ignore it of course. We may turn away into an optimism that holds for a while but wavers when the pressure of mortality becomes too great. We may choose to accept mortality and to justify that acceptance based on a conviction that it is "natural" to die or that death is not the end but merely a transition or even a new beginning. We find many routes out of the terror that death evokes in us. But death returns, and we are reminded of our mortality as we engage in life's unpleasant chores, those we tend to hide from others, like excreting. Indeed, if we could see through the skin to the organs and bones below, we would surely invent something like a skin as a curtain between us and the death that lurks inside us.

These are ways of toleration or accommodation. The core assumption they share is that death cannot be abolished. This core assumption has come under increasingly extensive scrutiny with an optimism spurred on by the technological revolution currently underway. Such optimism takes the renewed form of faith in technological transformation, albeit radicalized through the promise that we may be reborn literally as a new being no longer in thrall to death. We hear of the "singularity," not mere

J. Love and M. Meng, *Revolutionary Bio-politics from Fedorov to Mao*,
https://doi.org/10.1007/978-981-99-4745-4_1

enhancement of current capabilities but a transformation more extreme and far reaching than ever before contemplated in human history.[1] We become like gods, we turn an ancient fantasy or dream of immortality into a reality, though we cannot say what that reality might be since no one has ever overcome death. We stand before a revolution opening to us the vastest possibilities. Yet, our imagination fails because we can scarcely conceive of a being that is not subject to death. We are not revolutionary enough.

This book is about revolution. It is about what must be the most radical revolution, not a return to the roots, a beginning again, but a complete departure from roots and rootedness. This book is about the transformation of the rooted being, the human being fettered by time and place, into a being having no such restrictions, a being that has been a dream to some and a nightmare to others. This book is thus unabashedly speculative, though it examines a given line of thought anchored in an ancient tradition that has come to a certain peak in the modern era and has prompted various attempts to replace it, notably in twentieth-century Germany and China, with other revolutionary dreams, not of overcoming death but of finding peace or harmony with death or, if that is not possible, of finding consolation in a revolutionary politics "defined by continuous failure."[2]

Like all revolutions, the revolutionary dream to overcome death has a circuitous history. And although it stretches far back, perhaps originating with Plato, for the purposes of our narrative, it comes to fruition in the Soviet Union and ends, perhaps renewed, restored or discarded, in

[1] Ray Kurzweil, *The Singularity Is Near* (London: Penguin, 2005); see several articles by Michael E. Zimmerman: "Religious Motifs in Technological Posthumanism," *Western Humanities Review* 3 (2009): 67–83; "*The Singularity*: A Crucial Phase in Divine Self-Actualization?," *Cosmos and History: The Journal of Natural and Social Philosophy* 4, nos. 1–2 (2008): 347–370; "Last Man or Overman? Transhuman Appropriations of a Nietzschean Theme," *The Hedgehog Review: Critical Reflections on Contemporary Culture* 13, no. 2 (2011): 31–44.

[2] The quote comes from Wang Hui, "The "Modern Prince" and the Revolutionary Personality," trans. David Ownby (https://www.readingthechinadream.com/wang-hui-revolutionary-personality.html). In this essay, Wang suggests that an essentially Maoist politics of perpetual struggle for victory that never comes promotes a noble myth of relief from despair through collective action.

contemporary Communist China.[3] The nightmare comes to full expression in another revolutionary state, Nationalist Socialist Germany, and remains with us in the complex texture of resistance that marks the rejection of revolution in favor of a "return" to prudent peace with nature, to our status as natural beings having to learn our place in the cosmic whole rather than exploiting a chance to modify or determine it. The tone of this book is polemical: we take seriously the revolutionary imagination that demands complete transformation, utopia, final freedom, not merely for the few, the mighty and the rich, but for all. We have no wish to dismiss that imagination merely because of its apparent failure in the twentieth century and thereby to add to the chorus of despair about modernity that has become increasingly commonplace over the past several decades.[4] In doing so, we also seek to recover some of the peculiar optimism that has now become a straw man of contemporary conservatives, whatever their self-proclaimed political orientation, who would prefer we retreat from modernity itself in an effort at salvation through the embrace of mortality, vulnerability, and, in all cases, suffering as inevitable and even salubrious.

1 Universal Resurrection

Our story of universal revolution begins in Russia with a discussion of deification and a peculiar brand of Christian philosophy. We say "peculiar" because the most radical figure in the Russian installment of our story is Nikolai Fedorovich Fedorov (1829–1903), an unusual thinker espousing an outrageous idea: universal resurrection. Fedorov conceives the most radical notion of revolution as the transformation of nature in accord with the demands of human immortality. But Fedorov does not stop there. If the "common cause" (общее дело) he advocates aims to unify human beings in the pursuit of immortality or the conquest of death through technology, the task remains unfinished with the actual physical salvation of only one generation. Fedorov claims in fact that the "common

[3] In the *Symposium*, Plato makes desire for immortality the foundation of his narrative about the seeker ascending from the material to the immaterial world beyond it in order to love "beauty in itself, absolute, pure, unmixed, not polluted by human flesh or colors or any other great nonsense of mortality." Plato, *Symposium*, trans. Alexander Nehamas and Paul Woodruff (Indianapolis, IN: Hackett, 1989), 59 (211e).

[4] For a brief survey of contemporary expressions of despair, see Adam Kirsch, *The Revolt Against Humanity: Imagining A Future Without Us* (New York: Columbia Global Reports, 2023).

cause" is not so limited but involves a *moral* obligation to undertake the resurrection of all the generations who have ever lived on this earth. For Fedorov the task of resurrection so understood is intrinsically universal and, as such, the highest moral objective of all humanity as the species distinctively aware of death.

Not unsurprisingly, Fedorov's common cause has been met with skepticism, if not outright derision, by many commentators.[5] Yet, this was hardly the case in late nineteenth-century Russia. Fedorov managed to attract the attention of leading figures in the rich Russian culture of his day: he could count among his admirers Fyodor Dostoevsky, Lev Tolstoy and Vladimir Soloviev, Russia's leading nineteenth-century philosopher and an important thinker of deification whom we discuss at length in chapter one. Fedorov also attracted attention from some of the most adventurous minds of a younger generation, including the father of the Soviet space program, Konstantin Tsiolkowsky, and Vladimir Vernadsky whose notion of the world as both a bio- and noosphere has been influential.[6] Moreover, Fedorov's impact on the Russian revolution is itself not to be ignored; indeed, it is our intention to consider the revolution as an expression not only of Marxist ideas but more broadly of Fedorov's as well, an interpretation affirmed eloquently by the Russian novelist Vladimir Sharov. Of the Bolsheviks he writes,

> Neither then nor afterwards were they Marxist. They viewed Marx with irony and took no pains to hide the fact. His vision of socialism and communism; the role of the working class in history; the fact that the

[5] For example, Robert Bird refers to Nikolai Fedorov's "apocalyptic ravings" in his short book on Dostoevsky, and this phrase captures well the attitude of incredulity towards Fedorov held by many scholars. See Robert Bird, *Dostoevsky* (London: Reaktion Books, 2012), 183. But the most delightfully convinced opponent of Fedorov is John Gray who ridicules scientific optimism and immortalist hopes in his book *The Immortalization Commission* (London: Penguin, 2011).

[6] Fedorov is a key figure in what is referred to as "cosmism," a heterogeneous movement that sought to transform the world along utopian lines. See George Young, *The Russian Cosmists* (Oxford: Oxford University Press, 2012); Boris Groys, ed., *Russian Cosmism* (Cambridge, MA: The MIT Press, 2018); Irene Masing-Delic, *Abolishing Death: A Salvation Myth of Russian Twentieth-Century Literature* (Stanford, CA: Stanford University Press, 1992); Teresa Obolevitch, *Faith and Science in Russian Religious Thought* (Oxford: Oxford University Press, 2019), 145–155. As to recent history, see Anya Bernstein, *The Future of Immortality: Remaking Life and Death in Contemporary Russia* (Princeton, NJ: Princeton University Press, 2019).

> Revolution would begin not in Russia, but in one of the most advanced countries of the West; and above all, his notion of the meaning and purpose of man himself—all this struck them as quite astonishingly naïve. But Lenin impressed them as a man who could take decisions and get things done, and they had no doubt that in the end he would abandon Marx and cross over to Fyodorov. That time had not yet come, and this was no fault of Lenin's; it was simply that the nations of the world were not yet ready to hear the Word of Fyodorov, whose revelation was given to man too soon. For now it would survive and be preserved only as a secret, esoteric doctrine. Marxism would become the shell that saved it.[7]

The historical accuracy of this view is no doubt questionable. Yet, as an approach to the most far-reaching aspects of the Russian revolution, indeed of any revolution at all, Sharov's words can hardly be bettered. For the revolutionary impulse Sharov identifies rests on a fundamental assertion: that revolution must insist on complete transformation if it is to succeed. Otherwise, the revolutionary impulse ends up merely in a "turning of the page," merely in opening up another stage in history that leads to yet another and so on, in repetition that is infinite, the dreaded eternal return of the same. Fedorov's revolution is far more extreme; it is one that ushers in the end of all human striving in a final, changeless utopia—the end of history for Fedorov is the consequence of universal resurrection whose unspoken core is the conclusive suppression of suffering.

Lest one object that this has nothing to do with Marxism, we think it useful to recall a conversation between two robustly Marxist thinkers, Ernst Bloch and Theodor Adorno, precisely concerning death and the fate of revolutionary hopes which do not extend to the eradication of death. In reply to Bloch's connection of utopian freedom with the overcoming of death, Adorno says:

[7] Vladimir Sharov, *Before and During*, trans. Oliver Ready (Sawtry: Dedalus, 2014), 224. That numerous Bolsheviks and revolutionary figures, like Alexander Bogdanov and Andrei Platonov, had a serious interest in Fedorov is no secret. See Nikolai Krementsov, *Revolutionary Experiments. The Quest for Immortality in Bolshevik Science and Fiction* (Oxford: Oxford University Press, 2013); Nikolai Krementsov and Yvonne Howell, eds., *The Art and Science of Making the New Man in Early 20th-Century Russia* (London: Bloomsbury, 2021).

> [I]t moved me very much, Ernst, that *you* were the one who touched on this, for my own thinking has been circling around this point in recent times — that the question about the elimination of death is indeed the crucial point. This is the heart of the matter. ... I believe that without the notion of an unfettered life, freed from death, the idea of utopia, the idea of *the* utopia, *cannot* even be thought at all.[8]

Hence, we may argue that Sharov's point is indeed not far off the mark. Doubt and ridicule are of course always handy, but both Bloch and Adorno are, for their part, as serious as Fedorov, and their discussion is doubly relevant to our book because of the central role played in it by Martin Heidegger and his philosophy, according to Bloch, of "pro-fascist nihilism."

Although we may not endorse Bloch's characterization of Heidegger's philosophy, we do not ignore it either. In this respect, we recognize in Heidegger a comprehensive and sustained argument against the revolutionary project we sketch out in connection with Fedorov. Heidegger is the counter-revolutionary par excellence, the "conservative" who strives to undermine the utopian project (expressed, for us, most simply and radically by Fedorov) by casting all of Western history as a history of nihilism that terminates in the reign of technology from which only another God can save us.[9] If for Fedorov technology is the means to salvation, for Heidegger it is precisely the opposite—a means to the complete annihilation of human beings for which death is the central component of human identity, the expressive imperative of distinctively human existence.[10]

[8] "Something's Missing: A Discussion between Ernst Bloch and Theodor W. Adorno on the Contradictions of Utopian Longing," in Ernst Bloch, *The Utopian Function of Art and Literature: Selected Essays*, trans. Jack Zipes and Frank Mecklenburg (Cambridge, MA: The MIT Press, 1988), 7–8 and 10.

[9] We put "conservative" in scare quotes not least because Heidegger sees himself as engaged in a revolutionary struggle to overcome the metaphysical desire for conservation, stasis, permanency that he identifies as the goal of the modern, technological revolution. As Heidegger writes: "Thus the primordial and genuine relation to the beginning is the revolutionary, which, through the upheaval of the habitual, once again liberates the hidden law of the beginning," (Martin Heidegger, *Basic Questions of Philosophy*, trans. Richard Rojcewicz and André Schuwer (Bloomington: Indiana University Press, 1994), 35. Translation modified.

[10] What might account for such different views of technological revolution? The disagreement may partly come down to the question of the "goodness" of life or, as Jean-Jacques Rousseau puts it, "the sweet sentiment of existence." Is life sweet? If so,

Having set up the issue as a comparison of disparate, opposing views, we turn to China to investigate possible responses to it, and we do so for several reasons. While there is a historical connection between Fedorov and China,[11] and Heidegger's popularity in China continues to reach new heights, we look behind these matters of influence to consider the question of technological development and the creation of utopia as emerging more broadly from the revolutionary aspirations of the Soviet Union adopted and yet thoroughly reconceived in the Chinese context. Indeed, the adoption of Soviet and Western ideas in China seems to amount more to a transformation of them than reverent application. It is no secret that Mao did not simply accept the Soviet "experiment" as a model but strove to create a state that could fit the Chinese context without losing the utopian character of that experiment.[12] What emerges, and is in the process of emerging at the present time, is a reflection on the relation

why? We might identify three basic answers to these questions. One answer says that life may not be sweet given the unending spectacle of ugliness, crime, and stupidity that defines human history with its parade of tyrants, criminals, and fools inflicting suffering and death on others. This answer, seeing life as bad because it is created by a monster who makes us suffer and die, comes to the conclusion that it is best not to have been born. A second answer recognizes the centrality of suffering and pain to human life but suggests that these ills are essential to humanity from which they cannot be separated and for which bold affirmation is the proper response, the bold affirmation of finitude. According to this view, the sweetness of life lies in the struggle to come to terms with transience, a heroic and noble struggle to affirm not only that we are born to die but that to be born to die is good. A third answer does not deny the monstrous reality of suffering and death, nor does it deny that the creation is imperfect, but believes that its imperfections can and should be corrected through science and technology because life is good and should be infinite, not finite. This view expresses firm belief in the power of reason to save us: "The philosophers have only *interpreted* the world in various ways; the point is to *change* it." It also rejects the affirmation of death as pointless and cruel—universal salvation seeks to establish peace in the world and even beyond it (as Fedorov's disciples were eager to do). For them to see affirmation in struggle itself could only be a sign of despair.

[11] See Young, 201–207.

[12] We might say, borrowing from Li Zehou, that Mao planted and nourished the "Western root" of Marxism in such a manner that it could flourish in accordance with Chinese traditions. Li's slogan, "Western root with Chinese application (*xiti zhongyong* 西體中用)," views Marxism as the Western tradition most suitable to the Chinese emphasis on worldly pleasures and joys. Lest Li's interpretation of Marxism appear idiosyncratic, we might recall Leo Strauss's interpretation of modernity, in both its capitalist and socialist forms, as "political hedonism." Li Zehou, *A History of Classical Chinese Thought*, trans. Andrew Lambert (New York: Routledge, 2020), xiv and 327; Leo Strauss, *Natural Right and History* (Chicago: The University of Chicago Press, 1965), 169.

of the opposed positions we have described that seeks a way of reconciliation between total revolution and the supposed nihilism to which it might lead. Contemporary Chinese thought strikes a more conservative stance in the sense that the utopian state it conceives is not free of contradiction, nor does the Maoist tradition from which it stems and to which it responds insist on such freedom. If anything, the utopian state Mao proposed was one that attempted to maximize the benefits of technological development and transformation without seeking to achieve the absolute reshaping of the human envisaged by the most radical Fedorovian thinkers and artists of the Soviet state.

As Alain Badiou has suggested, to the puzzlement of many, Mao thinks in an almost "infinite way," for Mao extends revolution infinitely towards a goal that orders all striving though it may not be met.[13] We may argue that there is a Kantian aspect to this thinking which looks upon the revolutionary state as attempting to achieve infinite perfection. Yet, perfection remains the principal aim of the revolutionary state that in this respect marshals the energies of Heideggerian struggle not towards the re-affirmation of that struggle in an infinitely finite loop but rather with a view towards its elimination in continued progress towards ever more complete expressions of infinity and universality, which, in contrast to Soviet modernization, may not be in conflict with nature as the enemy to be liquidated. Instead, Mao's thinking insists on accommodation with nature, not hostility against it.[14] In this respect, he combines Marxism with Chinese wisdom, and Chinese wisdom, so we argue in an account of Chinese thinking on technology beyond Mao, affirms an attitude to technology and progress that seeks perfection in *harmony* and that views struggle not as an end in itself but rather as a struggle oriented towards harmony. We examine two important thinkers, one an implacable opponent of Mao, Mou Zongsan (牟宋三) and Yuk Hui (许煜), who develop the outlines of an attitude to technology that seeks harmony and balance, not Fedorov's radical salvation from death or Heidegger's rejection of such salvation as nihilistic, but the quest for salvation that includes all beings in harmonious acceptance of death.

[13] Interview with Alain Badiou, *Leap* 31, March, 2015.

[14] This claim may be overly generous since Mao also launched aggressive initiatives to modernize China, most notoriously in the Great Leap Forward that ended in disaster as had Stalin's collectivization drive in 1929. In neither case is there a discernible concern for creating a harmonious relation with nature.

By turning to China, our book reflects on the transposition of a largely Western conflict into the context of a culture that has completely different origins.[15] For the opposing positions toward revolution which we have set out have a clearly Western origin in the conflict between the consideration of human beings as natural beings no different from any other and as the crown of creation who may strive to transform themselves and nature thereby unifying heaven and earth. This characterization is of course hardly innocent of contention and stems from one of Fedorov's successors, Alexandre Kojève. Like Sharov, Kojève considers revolution as total and transformative, rejecting any so-called revolutionary impulse that is not devoted to the termination of the human being and therewith of history itself. Such revolutionary impulses, if not total, lead merely to repetition and thereby the assertion of finitude in a parade of ideologies that emerge and disappear in constant but predictable variation.

More comprehensively, by examining these broad currents of thought in the form of a philosophical narrative, our book reflects on some of the most basic questions that we can ask of ourselves: What is life? Why do we *want* to live? What is the goal of life, not just human life, but all life? Can we ever reconcile ourselves with suffering and death? Or do the pains of life so spoil the sweetness of life that we cannot resist denouncing or renouncing life?

2 The Infinite

These questions constitute the philosophical core of our narrative. While there are many popular and scholarly attempts to come to terms with technological revolution, they do not always deal adequately with the theological imaginary of the technological revolution and they rarely address squarely what seems to us to be a crucial issue: What kind of being shall emerge from this revolution? What kind of life shall this being

[15] In the Chinese tradition, for example, the kind of questions about origins that animate Western culture, questions about whence arises ignorance, error, evil, desire, etc., are simply not relevant, as Mou Zongsan notes in the following passage: "We do have a physical body, a sensibility, and personal desire; therefore, we have ignorance and moments of being dazed. This is as far as we can go in analyzing these kinds of questions. If people wish to pursue them further by asking why humans have sensibility or personal desire, then such irrelevant questions can only be answered with 'it is hard to know.'" Mou Zongsan, *Nineteen Lectures on Chinese Philosophy*, trans. Esther C. Su (San Jose, CA: Foundation for the Study of Chinese Philosophy and Culture, 2015), 310.

lead? What relation can such a being have to us, and the life we lead at present on this earth as we know it? Often the future life bears an uncanny resemblance to our present life, merely extended and eased. Such imaginings impose our present life on the future being, the finite on what is not recognizably finite. Yet, if this being is not subject to death, we have to ask in what way it can resemble our present humanity with our present life. If death is so terribly fundamental that we wish to eradicate it, then what exactly are we without death? If only individuals die, how can what does not die be individual?

We might usefully anticipate the properly philosophical argument of this book by turning once again to the infinite. We could argue that the revolution of which we speak in this book is in the final account a revolution that seeks to make of the finite human being an infinite one. For this is exactly what freedom from death must demand of us: no limit to our being can be countenanced. We become, in accordance with the most profound layers of Western thought, infinite. Merely to become infinite is remarkably perplexing, and we wish to give a preview of our argument more specifically with regard to the problem of infinity as *the* most pressing and promising theoretical problem of modern thought.

Alphonso Lingis puts it well:

> Europe—the spiritual destiny to be known as the West—was born the day that the idea of infinity emerged in human consciousness. It is infinity that the West pursues. Western cognitive life is not distinct from the wisdom of the East or that of the savage mind in that it holds to different concrete doctrines but in that it and it alone is in pursuit of an essentially infinite truth. Western praxis is distinctive in that it is not yoked to the satisfaction of needs, which are always finite, but, continually able as it is to produce new needs, it is able to give itself infinite tasks. Western ethical life is the production not of concrete types but of ideal norms. The West's God has infinity as his sole positively understood property; Western religion is the worship of the infinite. The Western expansionism, its imperialist ambitions, has its ratio not in climatic or biological factors that would explain the vigor of the European races, but in the spiritual format of Western culture which must transcend all simply finite goal-structures.
>
> The idea of infinity is not an idea of something, is not a concept, does not contain a content. Infinity is only as an idea or as an ideal; the idea of infinity is the idealizing form of ideas. Infinity cannot be given, actual; it can only be ad infinitum. It is only as aimed at; it indeed is the pure form

> of a telos. It is by becoming through and through aim, intentionality, that Western consciousness pursues infinity.[16]

This distinctive aim makes out of the finite human being an ideal being, an infinite being, a being not subject to the mere "satisfaction of needs." In case one may consider this empty exaggeration, let us think of the basic Kantian distinction between heteronomous and autonomous ways of living, the latter constituting an ethical ideal. Kant may have insisted that this ethical ideal was not realizable due to the stubborn persistence of individual inclination (as against universal duty); but it is only a brief step further to undermine Kant's caution by insisting on the most thoroughgoing project of autonomy that would require the eradication of individual inclination entirely as an unhelpful relic of finitude. And if Lingis persists in considering infinity as ideal, that is, as ideal because it cannot be realized, the radical core of modernity, as we see it, is precisely devoted to the realization of infinity, perhaps inconsistently in the case of Fedorov and arguably more consistently in terms of Mao as Marxist "modernizer."

For what would the new being be if it were a synthesis of the non-mortal and the mortal? In other words, if the transformation in being envisaged by an immortalist (let us use this term) were not complete such that the new being retained nothing of mortality, what could this new being be as a mere synthesis? It is clear that such a synthesis would not be adequate at all because no remnant of mortality can be tolerated by a being that overcomes mortality. If we regard finitude in the terms of Lingis as marked by needs (and wants), then the immortal or non-mortal being cannot be so marked—the immortal or non-mortal being simply cannot be finite in this way if, indeed, in any way.

There may be an objection based on interpretations of the infinite. Some could argue that the Nietzschean model of eternal return is a model of infinite life. The model is one whereby a finite set of elements is repeated infinitely: thus, the new being would in effect be different from mortal beings only insofar as it would not die but rather simply repeat a given narrative—a life—endlessly. This potential model of infinite life brings up important questions as to what is to be preserved in the infinite life. Could we choose a finite set of elements for repetition? Does

[16] Alphonso Lingis, *Deathbound Subjectivity* (Bloomington: University of Indiana Press, 1989), 13.

it even make sense to think along these lines—how does one overcome mortality by preserving basic features constitutive of it? Another version of this objection foregrounds the finite set of elements but not the notion of repetition. The new being as such would possess a finite set of characteristics that could interact with a putative environment in a broad number of ways. Yet, even here, the question of choice arises (and thus identity) as well as the broader, and perhaps thornier, question of environment—the mere need for an environment is a restriction that puts in question the durability of the new being: an infinite being would not have any needs or wants, thus no need of an environment.

The problems of synthesis are vast as anyone with even a passing knowledge of Christian thought may attest, for the most notable synthetic being is of course Christ. The question of Christ's two natures has long bedeviled Christian thought and led to some of the most acrimonious debates and violent heresies of the early church.

We mention the Christological issue in passing at the moment, though it looms large over this book. We may even argue that the "passion for infinity" Lingis mentions is not merely a philosophical passion but a deeply Christian one as well that prepares the revolution from finite to non-finite being. It is no accident that Fedorov was thoroughly immersed in theology and thought of himself as a Christian bringing to completion the project and promise of Christ. The theological dimension and origins of the modern technological revolution are of surpassing significance, and, as we will see, the problem of Christology arises anew in terms of the new being that may emerge from the technological revolution. Indeed, the problem of Christology may be among the most important in Western thought and in our current moment to the extent that technology seeks to achieve what Christian thought could not: the transformation of the finite human being into a god, an ill-defined god to be sure, just as the Christian God who hides in apophatic seclusion.

Viewed from this perspective, Fedorov may be considered less radical since his attempt at universal resurrection aims to make the finite life infinite: Fedorov seeks a transfiguration on the model of Mt. Tabor. Yet, it is a sharp question, as we have noted, whether the finite life can be rendered infinite without the loss of infinity—the basic Christological concern, the central tension that animates that concern, animates Fedorov's project of universal resurrection and tends to cloud its possibilities of success not only in a practical sense but in a conceptual one. Heidegger's perspective seems resolutely finitist or infinitist only in the sense of an eternal return of the same; and yet Heidegger's criticism of technological utopianism is squarely based on a rejection of eternal return because of the reductionism that is crucial to its success: it can only succeed by selecting what is to be made eternal. But what is to be made eternal? Is there an individual essence, algorithm, or "whatness"—a soul—in each of us to be immortalized?

The Maoist view finds a way that refuses merely to impose the finite on the infinite. Rather, the Maoist route is one of a striving to transform finitude that may never reach a final point or reaches it only by discarding or cancelling itself in a barely comprehensible process of transformation or, at least, one that we cannot grasp because there is a break or jump from the finite to infinite that undermines notions of continuity and process. No stranger to the continuum, Leibniz once questioned how one knows when there has been a transition from life to death, a transition that, in the sense of the infinite, is not unlike questioning what may be the highest number in an infinite set.[17] The problem here, as throughout, is how we may conceive the infinite in its own terms.

If the infinite is distinguished by its independence from needs or wants, it is indeed difficult to conceive of any infinite life. For finite life is essentially determined by needs and wants; the kinds of objects we consider as objects and their significance result from the needs and wants they both satisfy and expand. Even the notion, perhaps so obvious, of a discrete object, of the discrete itself, if determined by needs and wants, broadly drawn, as it appears to be, loses its significance in the infinite where the discrete dissolves: if every single thing is an infinity of things there are no things. In this respect, we are likely attuned to spaces and objects primarily by the ways in which they enhance or detract from our survival: the finite life is ruled, as Schopenhauer lamented, by the tyranny of the will to live. The infinite abolishes this tyranny, and, for an infinite being there is no given imposing itself on us, nothing "always already there," no need whose contours are fixed and reliable. The infinite being creates itself out of itself—it is perfectly spontaneous, independent and free.

One nagging question remains, hard to suppress: how is it that we can affirm the notion that the fulfillment of human freedom, as freedom from

[17] G.W. Leibniz, "Pacidius to Philalethes: A First Philosophy of Motion," in *The Labyrinth of the Continuum: Writings on the Continuum Problem, 1672–1686*, trans., ed., and intro. Richard T. W. Arthur (New Haven, CT: Yale University Press, 2001), 143–149.

the tyranny of death, and thus from the will to live, is to be found in our becoming infinite if seeking that fulfillment is itself ruled by the very will it seeks to overcome? How does one break free from the will to live while at the same time affirming it in arguably the most radical way possible? Perhaps only Heidegger heeds this peculiar logic, finding a more terrible servitude in the will to free ourselves from death than in death itself.

Ultimately, revolution is about hope for liberation from servitude; and yet, if that liberation imposes even greater servitude, the meliorist hope that feeds revolutionary striving may seem stuck in a dangerous dead end. If the Christological synthesis admits of no clear structural resolution that conclusively collapses the tension it expresses, we may well ask what sense inheres in finally overcoming that tension. The same may be said of maintaining it interminably: while finitude is hard to bear, it is also hard to overcome, being a burden that one is neither able to carry nor to discard to paraphrase both Voltaire and Nietzsche.

3 Chapter Outline

While cautious about adding the by now obligatory chapter description so as not to try the reader's patience, we think a brief sketch of the specific structure of the argument presented in this book is in order.

The first chapter provides an historical and conceptual orientation to the notion of θέωσις and Divine Humanity in the Russian context as a prolegomenon to an account of Fedorov's thought. Unfortunately, most of Fedorov's texts remain untranslated, so we confine our account to the few available English translations. The basic thrust of this chapter is to clarify exactly what Fedorov proposes in support of universal resurrection and several concerns about the conceptual coherence of Fedorov's proposals. These concerns offer a ready introduction to the second chapter that deals with Heidegger.

In our view, Heidegger provides an incisive critique of the immortalism advanced by Fedorov and his progeny and avatars. He does so by raising the question of the negative as the key sign both of finitude and individuality in human existence. Hence, our chapter examines Heidegger's account of the negative and consequently his concerns about the implications of creating a new being completely freed of the negative. Such a being cannot brook any limits, but, as such, can hardly be a being in any sense that is comprehensible to us as we are now. Moreover, such a being as one that cannot include or reflect any negativity cannot be individual.

Heidegger's critique resonates into our final section on contemporary Chinese thought. If Fedorov represents the extreme vanguard of immortalism and Heidegger the trenchant critique, we explore the possibility of a middle way. We do so by applying Badiou's interesting comment on Mao as a thinker as well as by exploring other strands of Chinese thought as expressed by Mou Zongsan and Yuk Hui. To what extent can one be an immortalist or infinite thinker without falling prey to the difficulties that emerge in Fedorov and receive extensive treatment in Heidegger? We build on the preceding chapters by considering the thought of the infinite in terms of the abolition of the negative and the retention of a notion of individuality—is infinite thought an exhortation to gain liberation at the cost of individuality or is it an exhortation for collective creation that does not seek to destroy, to rest purely in negation?

As an epilogue we consider this issue by returning to the Russian context and the trenchant, if implicit, critique of Fedorov that emerges from Alexandre Kojève's discussion of terror that runs almost wholly counter to Fedorov. Kojève views θέωσις as motivated by terror and hatred of the flesh, a corrosive self-hatred that undercuts the argument for immortality. Moreover, we suggest that the Chinese context connects our account to an exceptionally rich tradition through which to question some of the most basic assumptions of the modern, technological revolution that seems to be unfolding ever more quickly today with rapid developments in such areas as artificial intelligence. Indeed, the Chinese context offers a different notion of intelligence that questions the assumptions on which artificial intelligence rests.[18] Why create an alien intelligence potentially superior to ours? What is that superiority? Or more basically put: What is intelligence? Is there only one notion of intelligence, the mathematical-algorithmic intelligence of the Western mind?

4 Our Approach

As a final note, we want to address briefly our approach to the material covered in this book. This approach is in keeping with the notion of the infinite we have presented. Over the last seventy years, and further back, there has been a proliferation of historical narratives purporting to explain one or another aspect of modernity, its emergence and basic

[18] Yuk Hui, "On the Limit of Artificial Intelligence," *Philosophy Today* 65, no. 2 (2021): 339–357.

characteristics.[19] The style, if we may call it that, is often one of conceptual elucidation through a narrative approach that gives the genealogy of the concept and thereby supposedly a fuller understanding of it. The presupposition in this approach is that concepts have a history that is essential to them and which we must grasp in its fullness if we are to attain an adequate conceptual understanding. While this approach is very problematic when it comes to understanding certain (purely analytical) concepts, say, of mathematics, whose freedom from history is considered to be precisely their commanding power, it is much less so in the context of concepts developed in philosophy and theology where history clearly plays a fundamental role.

Now the problem that afflicts these kinds of narratives is twofold: (1) they have often tried to provide something like a definitive or "persuasive" account of a concept; and (2) there is simply a surfeit of them, each competing with the others, each claiming definitive authority, each failing to convince others of that authority, leading to further proliferation. Here the underlying conviction is that concepts have an essence that can be explicated appropriately through a narrative. While perhaps no one might openly admit to harboring such an "essentialist" view nowadays, few studies openly argue that the view they advance is just one among many with no pretense to define an essence and to provide lasting insight. The problem is delicate: on the one hand, narratives assert a certain authority merely by making a claim on our attention, while on the other hand, if we accept that discovering essences is elusive, we cannot avoid entertaining

[19] This need to explain modernity perhaps can be traced to Jean-Jacques Rousseau's *Second Discourse*. After Rousseau, Hegel, Marx, Nietzsche, and Heidegger offered histories of their own to supplant Rousseau's narrative of humankind's break from nature into civilization. More recently, Karl Löwith (secularization), Leo Strauss (ancient vs. modern), Hans Blumenberg (nominalism), Michel Foucault (governmentality, biopolitics), Jacques Derrida (logocentrism), Hannah Arendt (totalitarianism), Max Horkheimer and Theodor W. Adorno (enlightenment), John Milbank and Charles Taylor (secularism) have provided accounts of the modern era. See, for example: Hannah Arendt, *The Origins of Totalitarianism* (New York: Harcourt, 1968); Max Horkheimer and Theodor W. Adorno, *Dialectic of Enlightenment: Philosophical Fragments*, ed. Gunzelin Schmid Noerr, trans. Edmund Jephcott (Stanford, CA: Stanford University Press, 2002); Hans Blumenberg, *The Legitimacy of the Modern Age*, trans. Robert M. Wallace (Cambridge, MA: The MIT Press, 1985); Karl Löwith, *Meaning in History* (Chicago: The University of Chicago Press, 1957); Frederic Jameson, *The Ancients and the Postmoderns* (London: Verso, 2015); Leo Strauss, *Natural Right and History* (Chicago: The University of Chicago Press, 1999); Charles Taylor, *A Secular Age* (Cambridge, MA: Harvard University Press, 2007); John Milbank, *Theology and Social Theory*, 2nd ed. (London: Wiley-Blackwell, 2006).

doubt about the conclusions provided by any narrative. How are we then to evaluate a given narrative account?

In this regard, we turn to the striking notion of history that G. W. Leibniz sets out in *Discourse on Metaphysics* (1685–1686).[20] In the *Discourse* Leibniz takes a radical position concerning conceptual identity insofar as he applies conceptual identity to historical figures—he maintains that there are individual essences. What is more, Leibniz maintains that individual essences mirror the universe as a whole in the sense that to describe fully the concept of Caesar, as Leibniz discusses it in the *Discourse*, is to describe the entire universe, thus bringing together whole and part as equal.[21] To the extent we are unable to achieve this level of description (which is infinite according to Leibniz), we lack adequate knowledge of Caesar so as ever to define him properly.[22] Where history is involved, the problem of completion and thus of final knowledge may never be overcome, and this raises the question of whether, in the infinite, any historical account can claim to be more definitive or authoritative than any other.

The questions that follow from this conclusion are vast in scope and vexing. We seek only to emphasize one crucial point: that our study presents a specific approach to the concept of θέωσις that seeks to uncover its importance to the modern striving to overcome death through technology. Θέωσις is for us a central *means* or *entry point* into the labyrinth of intertwining currents that emerge in this striving and that are too often associated with "trans-humanism," "post-humanism" or biopolitics, however important their contributions to our perception of technological change and human transformation may be.[23] From the point of view of

[20] G. W. Leibniz, *Discourse on Metaphysics*, trans. Gonzalo Rodriguez-Pereyra (Oxford: Oxford University Press, 2020).

[21] Ibid., 17.

[22] Ibid., 18.

[23] These terms are quite confusing, and there has been debate about the distinction. Put generally and superficially, trans-humanism is oriented to human enhancement through technology whereas post-humanism looks to establish a thinking that is no longer based on narrowly anthropomorphic assumptions. The confusion arises from the convergence between the two positions in terms of the rejection of a certain model of the human in favor of another. The divergence comes from the suspicion that transhumanism is a radicalized expression of humanism that expresses the deepest currents in humanism. See, for example, Rosi Braidotti, *The Posthuman* (Cambridge: Polity Press, 2013) and Max More and Natasha Vita-More eds., *The Transhumanist Reader* (Oxford: Wiley, 2013). To

the tradition we examine θέωσις is the eminently human striving that has nothing to do with a "trans" or "post" situation nor ultimately with biopolitics, though political organization of a certain kind is obviously implied by the notion of revolution we invoke.[24] But this revolution is one that overcomes biopolitics through biopolitics just as it overcomes what is a bar to becoming truly human in the final perfection of our power to overcome ourselves. For this tradition, becoming truly human is precisely becoming God, and we are not truly human until we have attained that final stage of our journey. In this specific sense, we may say, with only partial irony, that we have never been human.

Our approach complements and offers a comment on others, but is in the end merely one. Nonetheless, we hope to set out a provocative account of modern technological striving that should encourage and engage different debates about the ends of technological striving; not to prove that our narrative has a certain authority, but to explore broader and different perspectives on the primary issues involved. The crucial point for us is to open up possibilities for looking differently at modern technological striving, for if all views are perspectives in an infinite continuum, as we may suggest, Nietzsche's words are quite appropriate:

> There is *only* a perspectival seeing, *only* a perspectival "knowing"; and *the more* affects we allow to speak about a matter, *the more* eyes, different eyes,

be sure, our book considers positions that may be included in the post- or trans-humanist debate, but we prefer to examine them free of the polemics and confusions associated with these terms to the extent it is possible to do so.

[24] The term biopolitics as referring to "life or biological politics," coming from Michel Foucault, might be a pleonasm insofar as it suggests a connection between politics and death already implied by Plato in the *Republic* (Πολιτεία). As is well known, the *Republic* opens with a discussion about the injustice of death by Cephalus, and the implication seems to be that the city offers a solution to the injustice of death, if ultimately an inadequate one. It is within this Platonic context that we understand biopolitics, and the revolutionary, Platonic dream to overcome the need for biopolitics once and for all. In contrast to Plato and his progeny such as Fedorov, Foucault does not imagine the possibility of overcoming biopolitics; rather, he limits himself to critiquing certain regimes of biopolitics that happen to perturb him: "I do not understand myself as a universal combatant for a humanity suffering in all of its different forms and aspects. … If I am fighting for such and such an issue, it is because, indeed, it is important to me in my subjectivity." Michel Foucault, *Wrong-Doing, Truth-Telling: The Function of Avowal in Justice*, eds. Fabienne Brion and Bernard E. Harcourt, trans. Stephen W. Sawyer (Chicago: The University of Chicago Press, 2014), 266.

> we know how to bring to bear on one and the same matter, that much more complete will our "concept" of this matter, our "objectivity" be.[25]

Our approach, in short, is to offer an imaginative or "speculative" narrative drawing on an arguably fundamental current in the western tradition with global import that seeks to provoke thought, not silence it, by claiming to produce an authoritative history. Unlike Fedorov, we cannot hope to resurrect every human being and context from the past, for such would be an infinite task whose outcome would be to overcome its finite origins. And here, at last, is where we see the main contribution of our book insofar as the most puzzling aspect of studies that address the end goal of the technological striving we investigate is a murky notion of what exactly the new being will be. If we express concern about the various labels that have been applied to designate this being as post- or trans-human, we also put the question squarely, albeit in a theological context, of what identity can be applied to this new being even in the theological context. Whether truly human or post- or trans-human, the new being, as a projection of our current imagination, remains so vague that we wonder if that new being is any different than the one imagined to exist in any of the many kinds of afterlife created throughout history.

This book, then, is in this sense also another inquiry into the utopian ideal—it is surely no accident that we mention Ernst Bloch and examine in some detail his nemesis, Martin Heidegger. We confront a harsh but insistent reality for we either attend to the creation of a new being, requiring a new political existence or none at all, or we face the prospect of incessant struggle without a goal, struggle that itself becomes a goal. If this model of struggle may have been adequate at some point, it is surely far from adequate in a world where we have created weapons, and are creating more, that achieve the universality sought by many utopias in the inverse form of total annihilation, not only of human beings, but of many others as well. For the parallel imaginary that imposes itself on us is of a leap into annihilation—the murkiness of the new being only serves to indicate to what extent the new being is supposed to shed qualities that define us as we are such that, if we are stripped of them, we have effectively terminated ourselves. In this regard, the post- and trans-human labels may reflect more honesty, because if becoming God

[25] Friedrich Nietzsche, *On the Genealogy of Morality*, trans. Maudemarie Clark and Alan J. Swenson (Indianapolis: Hackett, 1998), 85.

is indeed the most human striving, the end, a Christ reborn in God though very much in human form, cannot but remain mysterious to us. Again, we cannot avoid the problem: to become other, to pass from this side of the revolution to the other, seems equivalent to self-annihilation. The utopian impulse becomes not dystopian but an exoteric doctrine of self-annihilation opposed, however, to thinking that sees in struggle the proper end of human life, even to this day, despite the fact that, there too, annihilation in global wars is the ultimate prospect.

CHAPTER 2

Becoming God

Union with God is one thing, becoming God another. The Russian tradition is distinguished by its radicality: the aim of Christianity is salvation through becoming God: deification. Though this tradition looks back on a rich Greek patristic heritage, the emphasis on the essential homology of salvation and deification as literally overcoming physical death is one of the red threads running through the efflorescence of religious thought in the late imperial period, roughly between 1860 and the time of the Revolution. In Ruth Coates' apt words:

> By deification they understood becoming like God specifically in His immortality: overcoming physical death. Kärkkäinen's observation about Eastern theology, that 'it does not focus so much on guilt as on mortality as the main problem of humanity', is overwhelmingly true of these Russians, who look for nothing less than total ontological transformation, not only of themselves, but of the entire material universe.[1]

Becoming God or θέωσις, as it is traditionally referred to in Eastern Christianity, has a long history indeed. Norman Russell identifies three basic approaches in the patristic texts: nominal, analogical and metaphorical. If we may consider the metaphorical approach of most relevance in

[1] Ruth Coates, *Deification in Russian Religious Thought* (Oxford: Oxford University Press, 2019), 27.

J. Love and M. Meng, *Revolutionary Bio-politics from Fedorov to Mao*, https://doi.org/10.1007/978-981-99-4745-4_2

the Russian case, that classification remains essentially unclear—for what is it to think of deification merely as a metaphor or *façon de parler*? In this instance, deification seems little more than a way of directing action in this world, thus, as the metaphysically or ontologically transformative end to be pursued through right action, a kind of ethics. Now the problem with this thinking is clear: what exactly is the end toward which practice directs us? Russell writes:

> The metaphorical use is much more complex. It is characteristic of two distinct approaches, the ethical and the realistic. The ethical approach takes deification to be the attainment of likeness to God through ascetic and philosophical endeavor, believers reproducing some of the divine attributes in their own lives by imitation. Behind the use of this metaphor lies the model of *homoiosis*, or attaining *likeness* to God. The realistic approach assumes that human beings are in some sense transformed by deification. Behind the latter use lies the model of *methexis*, or *participation*, in God.[2]

Homoiosis and *methexis* presuppose very different attitudes to God; yet, they both maintain a basic ontological difference between human beings and God. Deification in this sense describes something like an infinite approach to God, an approach that cannot do more than engage in some form of progression towards an ultimately unattainable "object," or, as Alexis Torrance puts it, an approach that might be renamed "perpetually deferred deification."[3] Deification is, then, a primarily ethical ideal: it directs action toward what exists as an ever-vanishing goal. We use the term "vanishing" quite intentionally because the God toward which the ethical human being strives is distinctive; its being is unlike any other being of which we have experience. The likeness set out in Genesis 1:26 is a most unusual one.[4]

[2] Norman Russell, *The Doctrine of Deification in the Greek Patristic Tradition* (Oxford: Oxford University Press, 2004), 2. The literature in English is fairly extensive, exemplary is: Michael Christensen and Jeffery A. Wittung, eds., *Partakers of the Divine Nature* (Grand Rapids: Baker Academic, 2007).

[3] Alexis Torrance, *Human Perfection in Byzantine Theology: Attaining the Fullness of Christ* (Oxford: Oxford University Press, 2020), 2.

[4] The concern with "likeness" and "image" is fundamental. Important for our concerns is the distinction made by Vladimir Lossky: "The perfection of man does not consist in that which assimilates him to the whole of creation, but in that which distinguishes him from the created order and assimilates him to his Creator." Deification affirms man's connection to the creator and creative acting, not to the creation itself. See Vladimir

In the Eastern tradition the divine being is in itself totally elusive. The apophatic tradition creates a God whose being cannot be grasped in any way like a being we encounter in the world.[5] This God lies hidden and in his being completely inaccessible to the kinds of categories with which we classify objects encountered in the world. The sheer inaccessibility of God is a significant problem for deification: one cannot become, even through metaphor, what one cannot possibly know. The infinite approach to God is inevitably so if God is inaccessible or unknowable in his essence.

The identity of God is a constitutive problem for deification, though it seems rarely posed as such. No matter what approach one takes, one cannot reach what one cannot possibly know. Indeed, the more obvious question is how it might be possible to approach God at all if God is as elusive as the apophatic tradition makes God out to be. This God, as an identity, is ontologically distinct from all other things: this God is a pure "is" shorn of any predicates. Thus, the approach to God turns out to be an approach to this pure "is" that remains wholly undefined and paradoxical in the simple sense that the pure "is" remains outside any experience in the world we may have. But it also remains outside our imagination and science—for how can we know what has no predicates whatsoever? How can we know a thing or subject of predication that has more broadly none of the qualities we attribute to a thing, including the most basic, that it is in time and space?

Even the Platonic ideas that ground the original notions both of *homoiosis* and *methexis* have clear identities which relate to objects we can encounter in the world. The idea of the bed, so celebrated because of Plato's discussion in Book X of *Republic*, is clear because we have experience of the bed in the imitated form of the material bed in time and space. Indeed, we make the bed though we depend on an idea that we do not make in doing so. While grave problems arise when it comes to defining the relation between the immutable idea and its material instantiation in an object appearing in the inferior world of change and becoming, as noted in the *Parmenides*, the relatively simple, objective identity of an

Lossky, *The Mystical Theology of the Orthodox Church* (Cambridge: James Clarke & Co., 1991), 114.

[5] A crucial text in this regard is the *Mystical Theology* of Pseudo-Dionysius the Areopagite. See Daniel Jugrin, "Agnōsia: The Apophatic Experience of God in Dionysius the Areopagite," *Teologia* 67, no. 2 (2016): 102–115; Andrew Louth, *Denys the Areopagite* (London: Continuum, 1989), 78–98.

idea like that of the bed, ensures us that there is some tangible *thing* to which our attention may be directed.[6]

This is hardly the case with God for there cannot even be an idea of God. To the contrary, the Eastern tradition affirms a view of God that separates God himself from all appearances, a view whose essence is to preserve the elusiveness of God and not drag God down into identity with any one object or group of objects. Indeed, the apophatic way clears God of any predicates used to define an object in the world, whether really existing in a material sense or in the realm of imagination. If one were to imagine God, one would be doing so in error no matter where that imaginative product might be in relation to God—the notion that there are layers of being on the way to God allows for attributions of identity of some kind but not identification or full identification. And the elusiveness of full identification tends to undermine the coherence of partial identification as well.

And here the real sting of the apophatic tradition: if God is hidden from us, if the "is" of God is literally nothing for us, or indistinguishable from nothing, then we may say that becoming God in its most direct sense is becoming nothing: the apophatic way, as a way of *ethical practice*, must be a way of shedding predicates as we approach ever closer to God. By shedding predicates on our way to God we shed all identities that we hold and thus sever our relation in the end to time and space—we engage in a process of de-finitization or infinitization that renders us completely unfamiliar to ourselves, for even the self, as a product of predications, melts away. This utterly fantastic notion of deification is one of annihilation or, rather, self-annihilation on the way to the deity and thus resembles several traditions that prize annihilation as the proper end of human action.[7]

[6] See Plato, *Parmenides*, trans. R. E. Allan (New Haven, CT: Yale University Press, 1998), 7–8.

[7] We are thinking here of mystical traditions in particular, like that of Sufism where the concept of fanā (فناء) is significant as a kind of "death in life." See Henry Corbin, *Alone with the Alone* (Princeton, NJ: Princeton University Press, 1997), 119. Corbin refers to fanā as describing a return to oneself after dying away or "to endure after annulment." But the concept of "releasement" or *Gelassenheit* that Heidegger borrows from Meister Eckhart has a somewhat similar significance insofar as one continues to live while having left the self behind. Whether this kind of "death in life" is possible is a question to which we return at various points in the book.

Annihilation is manifestly not what is sought in deification, however. It is too caustic an irony to suggest that the striving to free ourselves of the bonds of mortality should lead us directly to mortality. We do not seek a complete end in a deceptive new beginning. This new beginning remains, however, most puzzling for the end of deification should not simply be an end. Is deification then a self-cancelling project as some might argue? If not, then how is it possible? For it must be coherent to be possible or to be considered as a possibility for human action?

The Russian response to these fundamental difficulties is remarkable because it weaves together doctrinal elements from Eastern Christianity with elements of German idealist philosophy in its endeavor to be holistic, universal and uncompromisingly radical, arguing for ontological transformation through divine humanity or literal resurrection as the only ways to attain genuine deification. We think it is instructive to look at this response from two perspectives, neither exclusively theological—those, respectively, of Vladimir Soloviev and Nikolai Fedorov. We begin with Soloviev who is especially sensitive to the problems we have canvassed; indeed, perhaps no thinker in the modern Russian context is quite so comprehensively sensitive to the problem of becoming God as Soloviev.[8]

1 Divine Humanity

Vladimir Soloviev (1853–1900) is a pivotal figure in nineteenth-century Russian thought. He is the creator of the fecund concept of divine humanity or Godmanhood (богочеловечество)[9] and his influence on a generation of Russian thinkers of all stripes was immense. His attempt to imagine a genuine fusion of God and man is instructive because it ultimately fails to constitute more than an infinite striving approaching

[8] We recognize that this statement, like similar statements, provokes opposition. Yet, Soloviev's seminal influence on Russian philosophy and Russian religious thought is quite hard to deny. Figures as diverse as Sergei Bulgakov and Alexandre Kojève were deeply influenced by Soloviev (and we mean influence perceived both positively and negatively), and the Nietzsche reception in Russia is unthinkable without him. We might go so far as to argue that Soloviev has not really been fully discovered as a thinker yet—some are truly born posthumously.

[9] Vladimir Solovyov, *Lectures on Divine Humanity*, trans. rev. and ed. by Boris Jakim (Hudson NY: Lindisfarne Press, 1995). We have also referred to the Russian text: Vladimir Soloviev, *Sobranie Sohineniia,* eds., S. M. Soloviev and E. L. Radlov, 2nd ed. 12 vols. (St. Petersburg: Prosveshchenie, 1911–1914), 3/27.

God, though Soloviev, perfectly aware of this difficulty, labors to avoid it. The criticism leveled at Soloviev is that, by maintaining the dyad of God and man, he maintains an infinite distance between them that he is unable to overcome.[10] We are not sure that this criticism is wholly fair to the extent that it underplays the innovativeness of Soloviev's attempt to bridge the distance between God and man. For if Soloviev has often been considered merely an assimilator of German idealism and, particularly, of F. W. J. Schelling, we detect more decisively the influence of Leibniz who perceives the infinite in every finite thing; but, more than that, we grant to Soloviev's thought far more originality and distinctiveness than the usual attributions of influence allow. Thus, we think a fresh approach to Soloviev, an expressly sympathetic one, is in order and appropriate at this point of our presentation because of the problems it articulates so clearly, primary among which is indeed that of imagining a being that is both infinite, or absolute, and conditioned by the everyday terms of finite human existence. That is: we return to the basic Christological problem, and Soloviev attempts nothing less than to resolve the Christological deadlock in a way we think is as close to Leibniz as it is to German idealism.

Soloviev's thinking involved quite a few turns in his relatively short life. We focus on the argument he sets out in the *Lectures on Divine Humanity* he delivered to a large audience in St. Petersburg in 1877–1878. He puts the basic question that guides the lectures early on, in Lecture Three: "The question is asked, What sort of *absolute* content can there be in life when it is a necessary natural *process*, conditioned on all sides, materially dependent, completely relative?" [Какое может быть *безусловное* содержание у жизни, когда она есть необходимый естественный *процесс*, со всех сторон обусловленный, материально зависимый, совершенно относительный?].[11] Perhaps the central, and most elusive, concept Soloviev tries to create is a viable concept of a conditioned absolute, a combination that would otherwise seem wholly impossible, something like a squared circle. The stakes are fundamental and Soloviev examines them in terms of two basic attitudes to the absolute, one negative, one positive. Soloviev recognizes that the apparently impossible combination of a conditioned absolute invites skeptical

[10] Alexandre Kojève, *The Religious Metaphysics of Vladimir Solovyov*, trans. Ilya Merlin and Mikhail Pozdniakov (Cham: Palgrave MacMillan, 2018).

[11] Solovyov, 25.

responses, and such responses coalesce in the notion of the negative absolute to which Soloviev devotes most of his attention at the beginning of the lectures.

The negative absolute is the absolute of self-erasure, the absolute achieved by the eradication of all marks of finitude through negation: it is the absolute, though Soloviev does not put the matter so bluntly, of suicide. The negative absolute can be realized only by elimination: for a finite creature the absolute is only possible in the negative, as the privation of existence because there can be no stable compatibility between the conditioned and the absolute. In Christological terms, the negative absolute is tantamount to heresy because it ends up denying the divine nature in the human being by indicating that the divine nature cannot be realized in a finite being or can be realized only, and this seems blackly humorous, by the termination of finite, conditioned existence.

Soloviev introduces the concept of the negative absolute initially as the failure to be content with any limit, as a rejection of limits:

> If human beings begin by confusing the absolute principle with the weak and beggarly elements of the world (and as finite natural beings we must begin with such a confusion), then to understand and realize that principle in its own actuality we must first separate it from, and oppose it to, those weak and beggarly elements of the world. To understand what the absolute principle is, we must first reject with will and thought what it is not. This absolute rejection of all finite, limited attributes is already a negative determination of the absolute principle itself. For a consciousness that does not yet possess that principle itself, such a negative determination is necessarily the first step toward positive knowledge of it. Contemporary consciousness, which has shifted the center of gravity from the absolute principle to conditioned nature, must go through a complete and decisive rejection of that nature to be able, once more, to apprehend the supernatural, absolute reality.[12]

The negative absolute results from a refusal to accept the conditioned as final—no point of view can be final or unconditioned, and thus no point of view can assert itself as being authoritative or absolute other than in the negative form of there being no absolute other than the absence of an absolute. The view Soloviev describes is one of endless critique, and it reflects the establishment of critique as the stand-in for the absolute:

[12] Ibid., 44.

critique becomes absolute in the sense that it cannot arrive at any point that it cannot surpass or view that it cannot refute or show to be wanting in absolute validity. The absolute has and can have no positive content. Rather, it is definitively negative, the rule (if that is the right word) of an absence or nothingness.

Soloviev connects this negative absolute with the flourishing of multiplicity and the particular as having absolute authority, for, when there is nothing that can withstand critique, there is nothing that can bind particulars together in a broader whole. The negative absolute thus leads to an atomization of particulars that in the process of self-assertion through critique end up dissolving themselves as particulars as well: where there is no unifying generality, there is also no particularity. The negative absolute excludes nothing and includes everything without providing support for any one position over another. The complete indifference of the negative absolute or its negatively egalitarian aspect—all views are equally correct and incorrect—invites inevitably contradictory self-assertion that can prevail only through the application of violence, sheer physical violence, that takes the place of authority. Such violence sets the stage for the chaotic war of all against all or for a world in which differences are resolved finally only through violence since there is no view that all must accept as absolute other than that the absolute is lacking. If the ground of all assertion is absent or a non-ground, any assertion is as compelling as any other.

For Soloviev, the negative absolute is only a moment in a greater development towards a positive absolute: "Not satisfied with any finite conditional content, humanity does, indeed, declare itself to be free from any internal limitation. It declares its negative absoluteness, which constitutes the surety of an infinite development. But the dissatisfaction with any finite content, with any partial, limited reality, is itself a demand for all of reality, for full content. And in the possession of all of reality, of the fullness of life, lies positive absoluteness."[13] The negative absolute restricts us to the realm of nature, of the natural particular, that seeks to attain satisfaction but can attain satisfaction only negatively in a self-assertion that resembles self-negation to the degree it refuses to be satisfied by any natural state and, in the final account, by nature itself, which it views as evil and dissatisfying. Soloviev speaks of the natural being because the

[13] Ibid., 17.

negative absolute allows for nothing to take hold of us other than nature. The negative absolute confers no positive identity, one circles in endless dissatisfaction, in the negative infinite that can terminate only in self-termination or dissolution. "In all its acts, the human will is a striving for natural existence, the assertion of oneself as a natural entity. The renunciation of that will is the renunciation of natural existence. But since nature is primordially given as all, since outside of nature nothing exists for human beings in the given state of their consciousness, the renunciation of natural existence is the renunciation of all existence. The striving for liberation from nature is a striving for self-annihilation. If nature is all, what is not nature is nothing."[14]

Soloviev associates the negative absolute with Buddhism as an initial step toward the positive absolute insofar as the negation of nature is the precondition to salvation in Buddhism, a salvation that is largely negative since it finds no other way out of infinite negation than dissolution in the creation of the human being without qualities or the apophatic human to whom any and no predicate is applicable.[15] Critique leads to the endless affirmation of failure, of dead ends such that critique cannot even ground itself and may turn to a critique of critique that succumbs to further critique without end other than in resignation or complete dissolution. Nature may remain, but we would not even be aware of nature as such since that awareness presupposes a position beyond nature that can sustain itself and does not collapse in the negativity of endless misprision.

Soloviev opposes the positive absolute to this negative absolute as its refinement. The positive absolute establishes a realm of awareness outside of nature that is stable. He associates Plato with the creation of this realm. The positive absolute as ideal and immutable. If the negative absolute dissolves itself, the positive absolute asserts itself in its otherness not only to the negative absolute but to nature as well. This opposition to nature is of considerable importance. Soloviev insists that nature in itself can produce only violence and despair as each particular being prizes its own above all others and cannot make sense of its own existence because

[14] Ibid., 42.

[15] If we may anticipate, Mou Zongsan, as we shall discuss in chapter four, views the Hīnayāna Buddhist orientation towards extinction as a lower, inadequate teaching from the perspective of the bodhisattva mind. The latter respects attachments but also transcends them so as to spread one's mind out to the world in ever-changing ways. The bodhisattva mind is therefore freed from the individual desire for either permanency or annihilation.

the capacity to undermine immediacy that is the hallmark of the negative absolute cannot find satisfaction. The positive absolute, on the other hand, arises from the projection of a stable and unifying universal reality that transforms nature—rather than repetition of failure *ad infinitum,* we can claim the universal ideas as unifying, transformative content.

Yet, the primary problem of the Platonic ideas for Soloviev lies in their ultimate inaccessibility. They lie in an ideal reality whose connection with natural reality is still largely negative. The emergence of the Christ-figure promises to transform this situation by expressly eliminating the distance between nature and the ideal realm. One need not eliminate nature in favor of an unattainable beyond because one realizes that nature and that beyond form part in fact of an original unity recovered in Christ:

> If the divine principle is to be everything for us, then anything that is not this principle must be acknowledged as nothing. If, as Christ said, we lose our soul to receive it again, then of course we also lose the world that we may receive it again. For, as we shall see, if the natural world considered in itself, outside of and alienated from the divine principle, is evil, deceit, and suffering, then, in its positive relation to, or considered from within, that absolute principle, the natural world becomes a necessary instrument or material for the complete actualization, the final realization, of the divine principle itself.[16]

This conventional pattern of loss and recovery governs Soloviev's narrative that, typically triadic, places the natural world at the basis of the ideal world that arises from it and with which it combines in the figure of Christ who seeks to unify nature and the ideal principle in the realization of the latter in the former. The realization of the divine principle in nature transforms nature and leads to the creation of the divine human, the example for which is Christ.

Soloviev lays the groundwork for the creation of a divine human in a very interesting way by setting up another triadic structure: "First, to be the foundations of reality, the essences sought must be indivisible units, real centers of being, i.e., atoms. Second, to produce the actual diversity of being, these central units must act and receive action, i.e., must interact with one another. Consequently, they must be acting or living forces — monads. Finally, for them to constitute the essential all, to be

[16] Ibid., 44.

the content of the absolute principle, these individual forces must represent a certain content, or be determinate ideas."[17] This structure consists of concepts derived from three philosophic tendencies that affirm the infinite, atoms forming the bedrock of Epicurean nature theory, monads for Leibniz and ideas for Plato. If the significance of atoms is plain within Soloviev's thinking of nature, Soloviev employs monads and the ideas to create the basis for the Christological synthesis reflected in the divine human.

The monad is the cornerstone of this thinking. Soloviev takes advantage of the important notion that each monad is distinctive and individual (an atom) while at the same time "a living mirror expressing the entire infinite universe in an infinity of ways."[18] The monad thus brings together the infinite universe and the monad, as an individuum, in a combination that seeks to combine the many and the one. If God is this infinite universe, the divine principle being infinite, the monad effects a harmonious synthesis of the individual, the one, and the many. The monad contains all within itself and is an integral part of the universe as well—the moment of reflection that brings the monad to itself is thus a crucial moment of discovery whereby the individual comes to realize its own

[17] Ibid., 55.

[18] Ohad Nachtomy, *Living Mirrors: Infinity, Unity, and Life in Leibniz's Philosophy* (Oxford: Oxford University Press, 2019), 2. This image has a remarkable resonance in Buddhist thought, the celebrated net of Indra so important to Huayan Buddhism that Mou Zongsan holds in high regard as close to the perfect doctrine he associates with another school of Chinese Buddhism, Tiantai. Here is the image as described by Francis Cook:

> Far away in the heavenly abode of the great god Indra, there is a wonderful net which has been hung by some cunning artificer in such a manner that it stretches out infinitely in all directions. In accordance with the extravagant tastes of deities, the artificer has hung a single glittering Jewel in each "eye" of the net, and since the net itself is infinite in dimension, the jewels are infinite in number. There hang the jewels, glittering like stars of the first magnitude, a wonderful sight to behold. If we now arbitrarily select one of these jewels for inspection and look closely at it, we will discover that in its polished surface there are reflected *all* the other jewels in the net, infinite in number. Not only that, but each of the jewels reflected in this one jewel is also reflecting all the other Jewels, so that there is an infinite reflecting process occurring (Francis H. Cook, *Hua-yen Buddhism: The Jewel Net of Indra* (University Park: The Pennsylvania State University Press, 1977), 2).

infinity and thus proceeds to cultivate the universe in itself as an idea, that is, as a direct intuition of the universe. The ladder of development Soloviev constructs in his account of this triad is one of progressive integration with the universe, in Soloviev's terms, the absolute, that is not a striving to achieve a new universality, but rather a return to a universality that had been forgotten through the fall.

Like most "golden age" narratives, including those of Plato, Soloviev's has the difficulty of explaining not only how but also why an original harmony has been lost. Soloviev is quite conventional in his deployment of the Christian narrative of fall and redemption to explain our ignorance of the absolute that lies within all of us. But the philosophical foundations of his account are more original since he relies on apparently disparate philosophical tendencies (Epicurean atomism and Platonic idealism, for example) to create a narrative that labors to address the fundamental Christological problem: how does one combine the infinite and unconditioned, the absolute, with the conditioned and finite? Soloviev's solution relies primarily on Leibniz for its structure and on German idealism and the Christian narrative for its exposition of that structure as a narrative of development, of egress and return or fall and redemption with the redemption coming not in an afterlife but in our life in this world.

Now, before going further, we should like to address an objection to Soloviev's project suggesting that his synthesis of individual and universe, finite and infinite, human being and God is impossible within his own terms. The basic gist of this objection is that Soloviev institutes a striving for the realization of the infinite in the individual that cannot work because that striving simply asserts the distinction between the individual and the infinite in the guise of God. Put differently, the need for striving itself asserts the distance between individual and God because the aim can never be assimilated since it is infinite. We end up with something akin to the Kantian striving to achieve the regulative ideal, in practical terms, in the sense of the universality commanded by the categorical imperative. For Kant this striving cannot achieve its end because it is necessarily impure given that our empirical being and its inclinations can only be eradicated in universality with the termination of those inclinations and, thus, of the empirical being as well. Soloviev, likewise, merely institutes an infinite striving that returns again and again to the limited individuality it seeks to overcome through an infinite task of self-discovery. Far from creating the basis for a divine human, Soloviev merely reaffirms the finite humanity of that being and its infinite distance from divinity. He offers us a version of Kantian "skepticism" because the striving cannot achieve the goal it sets out to achieve, casting light on the futility that attaches to the regulative ideal.

We may respond to this objection by claiming that it misses the Leibnizian dimension of Soloviev's thinking. Let us return to the "fantastic" idea of the monad as "living mirror." The monad contains everything in itself:

> This is a very simple idea. The notion of the subject contains everything that happens to a subject, that is, everything that is said about the subject with truth. So, "Adam sinned", sinned at a particular moment belongs to the notion of Adam. That gives one pause, right? Crossing the Rubicon belongs to the notion of Caesar, very good. I would say that there, Leibniz proposes one of his first great concepts, the concept of inherence. Everything that is said with truth about something is inherent to the notion of this something. So, this is the first aspect of sufficient reason, it's the development of sufficient reason. Fine. Only here we are, we can no longer stop, and when we say that, listen to me closely.[19]

Deleuze alludes to the key point: a subject, or living monad, is the iteration of its predicates expanding "outward" to include everything in the entire universe.

> So, you see that Leibniz cannot – he cannot, he has no choice here – he cannot fall back on the solution: since every true proposition is analytical; the world is thus contained in a single and same subject which would be a universal subject. He cannot since his principle of sufficient reason implied that what was contained in a subject – thus what was true, what was attributable to a subject – was contained in a subject as an individual subject. [*Pause*] So, he cannot give himself a kind of universal mind. He has to remain fixed on the singularity, on the individual as such. And in fact, this will be one of the truly original points for Leibniz; this is the

[19] Gilles Deleuze, *Seminar on Leibniz: Philosophy and the Creation of Concepts*, trans. Charles J. Stivale, 12, published on *The Deleuze Seminars*, a Purdue University website devoted to Deleuze's seminars.

> perpetual expression in his works: substance – for him, there's no difference between substance and subject for him; for other philosophers, there is a difference, but for him, there is a difference, but for him there is none—substance is individual.[20]

There is no isolated particular. Rather, all individuals mirror the universe completely and thus the divine principle is at work in them to the extent that they open themselves up to the universe as a whole rather than insisting on remaining closed off from the whole in their particularity. The insistence on particularity is the natural attitude or the attitude that remains at the level of nature distant from the universe and, to the extent of its insistence, maintaining that distance. Soloviev, not surprisingly, characterizes this insistence as evil because it rejects the divine principle and any community with other beings, all of which are seen exclusively from the natural attitude in terms of their utility for the particular being—nature as the realm of violence and predation turns its back on the whole. Now, to do so is not necessarily evil if one is wholly absorbed in nature, but it is clearly an ingredient of Soloviev's thinking that human beings are not so absorbed—indeed, they are never fully absorbed in nature. As a result, the refusal to leave the natural attitude is a refusal to be human as the being that recognizes itself only negatively in nature and positively in its transcendence of nature.

When we speak of transcendence, we mean precisely the transformative recognition of the infinite in each particular as well as what Soloviev refers to as the realization of the idea in each individual that, as constituting a unique perspective in the infinite universe or absolute, is forever distinct, individual, having an individual idea predicated on its perspective. If all beings are monads that reflect the infinite universe or absolute, they each do so from a particular point that only they express and whose expression is necessary to the expression of the infinite whole, as necessary as any other "infinitesimal."

We may add, however, another caution stemming from Leibniz as well. For Leibniz divides truths into two categories, two kinds of inclusion of predicates in a subject, a given being: (1) analytic truths whose contrary involves contradiction and (2) truths of fact (*veritates facti*) involving facts which could have been other, whose contrary was possible—these truths require infinite affirmation. This is so because the rules that may apply to

[20] Ibid., 16.

these facts are not denumerable in advance, as is the case with analytic truths where the rules establishing the truth are clearly finite. Hence, the problem of the infinite in its negative sense returns with truths of fact: these would apply to persons, in particular, for one of Leibniz's innovations is to give detailed conceptual content to individual persons or proper names. This is a logic ostensibly applicable to the particular, as such, and not merely as an element belonging to a category. Here we return to the notion of Caesar that Leibniz mentions in *Discourse on Metaphysics*. The notion "Caesar" must include everything that happens to Caesar and, as a mirror of the infinite, everything connected with what happens to Caesar and so on *ad infinitum*. The union of particular and the general or universal demands that the universal appears in every particular and that every particular is the universal. Caesar is Caesar because of the structure of events that define or condition his position in the universe, but he is always also a mirror of the whole universe from a given perspective. There is thus a tension between the notion of universal rules defining all beings and an individuality that cannot be grasped by universal rules or, at least, by a finite set of universal rules that always end up by working categorically as applying to a given type of object. While nothing precludes our considering an infinite set of universal rules, the question remains as to how an infinite set of universal rules could constitute knowledge for us because we could never master them all nor free ourselves of a vitiating, infinite categorization, whatever that might be. Hence, an infinite set of rules undermines the possibility of knowledge of a particular in all its particularity thereby revealing a gap between human rule-based or categorial comprehension and a putatively divine comprehension—intuitive and immediate—that cannot be bridged.

Returning to Soloviev, we recognize a similar tension and have thus finally to decide whether the critique of Soloviev is justified insofar as the monad's process of self-discovery—that is also a discovery of the universe and the divine principle at work within it—does not merely bring us to an unending, if not Sisyphean, striving, the problem of reconciling the absolute with contingent existence remaining perpetually unresolved. Put simply: How can one truly become divine if the divine principle is always beyond the reach of a finite, contingent being? Soloviev:

> In the pre-Christian course of history, human nature, or the natural human element, was the basis or matter, while the divine mind (*ho logos tou theou*) was the active and formative principle. The result was the God-man, that is,

> God who has received human nature. In the same manner, in the process of Christianity, divine nature (the Word made flesh, or the body of Christ, Sophia) is the basis or matter, while human reason is the active and formative principle. The result is the man-god, that is, man who has received Divinity. And since a human being can receive Divinity only in his or her absolute wholeness, that is, in union with the all, the man-god is necessarily collective and universal, that is, is all-humankind, the Universal Church. The God-man is individual whereas the man-god is universal.[21]

How does an individual become universal, the man-god (богочеловек)? We have discussed the essentially Leibnizian idea of the monad coming to full knowledge of itself. Yet, if that knowledge is infinite and categorial, is it possible for a finite being to possess it, to be a "subject" that becomes substance, as Soloviev puts it otherwise, quoting G. W. F. Hegel from the Preface to the *Phenomenology of Spirit*? The questions do not cease. And they do not cease because it is difficult to grasp how an individual subject, rooted in time and space, in a specific perspective, can become universal or absolute. And, even if it could become universal or absolute, what would such a life be? Hence, the skeptical position remains, unrefuted. Another way of looking at Soloviev's claim, again from the Leibnizian point of view, is to suggest that the incorporation of the infinite is something more than the discovery by the monad of all the rules that undergird the formation of the infinite universe. Universality is acquired not by the acquisition of finite or even infinite rules but by something akin to immediate intuition—the monad sees itself somehow fully reflected in all other phenomena and, in so doing, absolves itself of its own finitude. Yet, in absence of such intuition, neither a finite nor infinite set of rules can grant to a monad perfect self-knowledge, it cannot ever attain perfect knowledge—something must always be missing.

Thus, we must be quite careful. What exactly is this absolution other than an impossible "knowledge beyond knowledge"? If an individual knows of the rules that shape its own formation, does that equate to divinity? Does that absolve the individual of finitude in the most direct and pressing sense, namely, death? God is undying as the referent we use, the shorthand, for the eternal rules or principles that govern all

[21] Solovyov, 173. Mou Zongsan writes: "God is an infinite being, and an infinite being does not manifest particularity." See Mou, 4.

phenomena. Individuals die. How then can the acquisition of knowledge of these principles grant to us the same status as God? We are still the conditioned being lacking immediate or intuitive knowledge, whereas God is not: for God is what conditions possessing an immediate intuition that we cannot: knowing and creation are one and the same. There is simply no way to make the leap between conditioned knowledge—that pertaining to conditioned, created beings—and the immediate "knowing" that creates and governs them.

Government is an essential word in this context because God, or the divine principle, rules whereas the monads or individuals are ruled. The old distinction between God as the *ens increatum* and phenomena as *entia creata* returns to affirm an essential and unbridgeable difference. Soloviev emphasizes this difference with the fall and redemption narrative structure he deploys because only the ruled can have a history: the rulers themselves do not. Soloviev makes this clear by indicating that human beings depart from an essential harmony that they rejoin. Such a harmony cannot be disturbed in the harmony-creating principle itself. The affirmation of difference is a fundamental aspect of Soloviev's narrative, and it is quite difficult to grasp how that difference can simply be overcome if it is indeed essential to the narrative, the foundation of the narrative, the foundation of the understanding of God and humanity that is presented in the lectures. Indeed, we could argue that Soloviev's attempt to overcome that difference is belied constantly by his own exposition that ends up self-refuting.[22]

In any case, the attempt by the individual, the contingent being, to erase its contingency by accepting the divine principle is unconvincing and seems destined to failure based on the problematic assertion itself for true harmony cannot easily be grasped as leading to disharmony and then a return—such a narrative applies to, and affirms, contingent or creaturely nature. The pattern is an old one too, essential to Christian narrative,

[22] At play here is a question that arises in connection with Sophiology: Is there a univocal concept of being at work, in which case God is immanent in the creation, or a transcendental concept of being that typically involves a notion of God as transcendent? The univocal concept cannot explain effectively the difference between God and creatures based on the uncreated nature of God and the created nature of those creatures, while the transcendental notion has immense difficulty in explaining how creatures can have any relation at all with God. Both cases identify a key problem for deification narratives because the difference between God and man, as the uncreated to the created, strains the relation of likeness established in Genesis 1:26.

and no clearer than Christian narrative when it comes to "solving" the Christological dilemma that strives to combine what is inherently resistant to combination and to do so in a way that favors non-creaturely being. There is no clear way, in other words, of progressing to the divine without inviting self-eradication, a return to the very negative absolute that Soloviev works to overcome or a return to the distance between God and man that he seeks to eliminate. Hence, the promise of bringing divinity to earth ends up as an imposture: so long as we suffer and die, so long as we remain in the prison of the flesh (*sōma sēma*), we can only ever deify ourselves through metaphors of infinite striving, which, though perhaps palliative, call into question the soteriological promise of Christianity itself.[23] As Alexis Torrance asks:

> Is it a literal deification, and if so, how literal?; or is it simply a nominal deification, an imputation merely of the title 'god' or 'divine' to the Christian, which has little to no content? … To say that human deification is 'metaphorical' and never literal would be tantamount to saying that the Incarnation itself was 'metaphorical' and not literal. The doctrine of deification simply cannot be detached from the doctrine of the Incarnation: to pronounce on one is to impact the other.[24]

2 Universal Resurrection

Unsatisfied with θέωσις as an infinite striving that never reaches its goal, Nikolai Fedorov takes a complexly different course from Soloviev who referred to him as his "teacher and spiritual father." He combines the most aggressive self-assertion of human power and authority in the creation with an ethical doctrine that requires such aggressive self-assertion as a duty to all past generations whereby universal resurrection, the resurrection of all human beings that have ever existed is the ultimate goal. This "supra-moral" doctrine may do little to convince us that Fedorov is after anything more than the realization of radical egoism, hardly leavened by its collective aspect. Thus, Fedorov's ethical doctrine may fall flat with those who understand ethics as essentially devoid of

[23] Whether Christianity overcomes the Platonic problem of *methexis* is the key question here. See Boris Maslov, "The Limits of Platonism: Gregory of Nazianzus and the Invention of *theōsis*," *Greek, Roman, and Byzantine Studies* 52 (2012): 440–468.

[24] Torrance, 33.

self-assertion or egoism as Soloviev and most Christian and Post-Kantian thinkers seem to do. To the contrary, Fedorov is simple and direct, and there is no need in Fedorov's thinking for metaphor or imagination other than the technical imagination required to achieve two crucial events: (1) the creation of the technical means to preserve individual human lives permanently and (2) the application of that technology not only to one generation but to all generations in the human chain, from the beginning on.

As we have noted, such an extravagant way of thinking may seem quite absurd to most at first blush. But one has to keep in mind the extravagance of the imaginative world Fedorov inhabits as a self-avowed Christian thinker: What could be more extravagant than the notion of resurrection in the first place?[25] Is Christianity to be dismissed as absurd as well? Or is to be reduced (let us say for the sake of argument) to a metaphor, a fervent hope or ridiculous presupposition? To borrow a phrase from Dostoevsky, is Fedorov merely a ridiculous man? He surely was not taken to be such, and by some very sober minds like that of Leo Tolstoy who admired him greatly. In our view, it is important to examine Fedorov's project sympathetically in order to appreciate its logic, however fantastic some aspects of it may appear to us. Fedorov's extravagance may come to challenge the wisdom of our "sobriety" as it may also come to address some of the deep currents in the current technological revolution. For one of Fedorov's most intriguing features is his attempt to weave together, almost seamlessly, theology, philosophy and technology. His thinking reveals the theological roots of modern technology in sharp contrast to thinkers like Soloviev who refuse to go as far as Fedorov or end up, as so often in the theological tradition, with a claim for divine humanity that demands metaphor and thus admits its failure to make human beings divine in the only sense that counts for Fedorov: the literal sense of making human beings immortal and eliminating history or historical narrative—which itself is nothing more than a false hope of recuperation of the dead—in the resurrection of all those who have ever lived. If we look at Fedorov's project from this perspective, we see the boldest attempt to assert the Christian promise of divinity

[25] "The only—theistic—mistake of Christianity" in Kojève's words. See Alexandre Kojève, "Hegel, Marx and Christianity," trans. Hilail Gildin *Interpretation* 1, no. 1 (1970): 42.

that has ever been made and that boldness and extremity, if nothing else, should demand our attention.

Fedorov's longest work begins thus, quite suitably, with a discussion of technology, a particular technology useful for military, destructive aims but also, possibly, for helping to end a drought by regulating the weather. Fedorov asks why it is that technology, capable of being used for the purposes of the advancement of human life, becomes used in eliminating human life or causing the most terrible suffering. His point here is of the greatest simplicity: Why should technology be a tool of war at all? Should technology not be a tool establishing a peaceful and thriving society? Like many of Fedorov's questions, the simplicity reveals the depth of the problem; for Fedorov interrogates the propensity to violence that stains human history with blood and does so almost always as violence condemned by all, including, all too often, those who perpetrate it. The latter live under the banner of slogans of ancient provenance as to the need for a strong military presence to preserve the peace—*si vis pacem para bellum*. For Fedorov, these arguments are diabolical: they deter us from our destiny as shown by the figure of Christ and ensure that we do not devote our time and energy and intelligence to a common task that will unite us.

This common task (the "общее дело") is fundamental to Fedorov. And we may add that Fedorov's common task has important antecedents in more recent Western thought. Baruch Spinoza's political project is in some respects notably similar in terms of the unity it seeks to create, though Fedorov's monarchical politics and Christian orientation are far from Spinoza. The basic connection is the affirmation of politics as a project of inclusion that suppresses any and all aggression among human beings in a common project of advancement of human "perseverance in being," i.e. survival. The target of Fedorov's aggression, it is indeed fair to call it that, is nature, the "blind force" (слепая сила) that we have allowed to persist in dominating us. The clearest sign of nature's domination is of course death, and it is death that Fedorov seeks to eliminate once and for all.

Let us take a closer look at the basic propositions of Fedorov's thought. To do so, we shall provide a condensed account of Fedorov's longest and most sustained text.

On the Problem of Brotherhood or Kinship, on the Causes of the Non-brotherly, Non-Kindred, that is, the Non-peaceful State of the World, and of the means for the Restoration of Kinship: A Memorandum from the 'Unlearned' to the 'Learned', Clergy and Laity, Believers and Non-believers.[26] The unwieldly title of this 300-page text is emblematic of Fedorov's style in which he attempts to describe as fully as possible the common cause. The comprehensive style provides an intriguing counterpoint to similar styles in earlier Russian prose, in particular the celebrated плетение словес or "word weaving" that sought excess as a way of approaching the deity negatively, as describing in effect what the deity was not or, better, exceeded in its uncompromising *negativity*. Fedorov's style is a positive counterpoint to плетение словес because Fedorov endeavors to describe positively what is and could be. While this may seem a fanciful point, it is in fact of the utmost importance in Fedorov's thinking since Fedorov is seeking to overcome a negative approach to the deity, as infinitely distant, to a positive approach by which human beings become one with the deity assuming a positive and not a negative identity as resurrected, saved beings.

It is worth pausing for a moment to consider the boldness of Fedorov's common task. As we know, the dominant tradition in Eastern theology thinks of God largely in negative or *apophatic* terms. Perhaps the most influential single document establishing the basis for the apophatic attitude is the *Mystical Theology* of Pseudo-Dionysius the Areopagite, a strikingly short text given the long shadow it has cast across the theology of the Eastern church. God is, in a word, not even a bare "is" for He is what or "who is beyond all being and knowledge": God is truly alien. The radicality of this apophaticism has not always been noted because, typically, Pseudo-Dionysius is considered a Neo-Platonic thinker for whom God is the ineffable one, likewise placed beyond experience.[27] Yet, Pseudo-Dionysius' God is utterly and completely beyond all being and knowledge. There is nothing we can know of Him, not even that He, in the contradictory phrase, "is nothing." Hence, the concept of God that emerges from Pseudo-Dionysius is radical because it is completely empty: God is not reason, not intelligence, not "being," and not even

[26] We have used as a basic resource one of the few translations of Fedorov into English: N. F. Fedorov, *What Was Man Created for?*, trans. Elisabeth Koutaissoff and Marilyn Minto (Lausanne: Honeyglen Publishing, 1990). The authoritative Russian edition is in 5 volumes and edited by two well-known scholars Anastasia G. Gacheva and Svetlana G. Semonova. See Nikolai F. Fedorov, *Sobranie sochinenii v chetyrekh tomakh*, eds. A. G. Gacheva and S. G. Semenova, 5 vols. (Moscow: Progress, 1995).

[27] See Louth, 78–98.

"not-being" or nothing. This is a God which defeats any connection with the human mind and yet remains "there" in an evacuated space. Such a peculiar God cannot be reached in any way, and one has to wonder if such a God is any different from the *absence* of God. Such a God cannot offer any guidance for deification other than that deification would have to be a state of transformation beyond anything the human mind could think. The complaint might return that deification is an expression of death, of something like a death drive at the heart of θέωςις that is wholly counter to any concept of θέωςις as bringing human beings closer to God.

Fedorov takes another tack and focusses on Christ as the realization of human being as a being. For Fedorov realization of the being of Christ is most distinctly expressed in resurrection whereby the fragile finite being becomes durably infinite in its finitude. Certainly, infinite finitude must appear to be an unusual concept, for we may well ask: How can there be such a being as an infinitely finite being? A fully finite being would be, we presume, one unable even to conceive of itself as a being. A "perfectly" finite being cannot be self-aware.[28]An "imperfectly" finite being may be one that is aware of itself as finite and yet is unable to change that finitude. An imperfectly finite being is simply condemned to acquiescence in finitude. The common consideration of self-awareness or "self-consciousness" as a disease appears to come from this perception of imperfection that arises from the awareness of oneself as finite. To "perfect" the "imperfectly" finite being in Fedorov's terms is to make it perfect by eradicating the defect, but not in the most commonly advised way, that of freeing oneself of self-awareness; or of giving oneself up to palliative illusion as Nietzsche might advise us for whom Greek superficiality and "lightness" is a product of depth.[29] Fedorov rejects both of these options and the conviction they express that our imperfection, as imperfectly finite beings, does not consist in that we are conscious of our finitude but that we are simply finite! He thus argues quite logically, from this premise, that the only way to repair this imperfection is the one revealed to us by Christ: the perfection of human being becomes the eradication of finitude such that our self-awareness no longer must be of a kind of unconscious mourning but can rejoice in infinite life. Now this

[28] This claim is provocative in the sense that it presupposes the possibility of a "perfectly" finite being.

[29] Friedrich Nietzsche, *The Gay Science*, ed. Bernard Williams and trans. Josephine Nauckhoff (Cambridge: Cambridge University Press, 2001), 9.

infinite life is the problem because it is not really infinite life that Fedorov seeks. Indeed, it is not clear what infinite life may mean in any case. What Fedorov pursues is infinite finitude, as we suggested, or, in other words, the infinite sustainment of a finite being. We must have due care as to what Fedorov hopes to attain by returning to the basic question of what an infinite finitude can be, and the nearest solution we can find is akin to Nietzsche's eternal return or a cyclical model of being whereby the finite being lives in finite cycles that are repeated infinitely.

We are not convinced that any other option is available. If Fedorov seeks to achieve the perfection of human life by eradicating its finitude while retaining a human identity, we think it is doubtful that there can be any model of such a life other than the rather ancient cyclical model—and even here, the question of self-awareness cannot easily be addressed, for not only all actions of the being in question must be repeated and repeatable but so must all its attitudes to those actions. By ensuring that self-awareness is thus assimilated to the structure of repetition applicable to the being's constitution and actions, self-awareness is itself reified and becomes as unfree and necessary and finite as the actions themselves. Simply: self-awareness as it is dissolves and is in effect eradicated by repetition.

The other chief choice is that of the infinitely infinite being. If the perfectly finite being is not human, the imperfectly finite being either finite or infinitely finite, the other choice must be an infinitely infinite being. Such a being, as we shall see, is barely distinguishable from the God of pseudo-Dionysius. The infinitely infinite being as human has to arise from two contrary assumptions: (1) either the human being is always already infinite; or (2) that the "infinitizing" of the human being brings forth a transformation whereby the human being becomes infinite.

If the human being is always already infinite—what can such a being possibly be? We know that such a being can be if not infinitely at least indefinitely repeating in its basic species-being. Yet, it is manifestly not species-being that Fedorov seeks to resurrect—there would be no sense in doing so because it is not the species that dies but rather the individual. Hence, Fedorov is strictly focused on the resurrection of the individual. We have already briefly examined this resurrection as infinite finitude, but here we are exploring another possibility that may bring us to Leibniz once again because Leibniz can give us a good account of an infinite individual, a radical concept. The concept is radical because Leibniz locates infinity in all individuals and individuality is defined as a specific "reflection point" on the infinite, as a specific point that itself can be broken down into further points if indeed all organic beings are infinite machines made of infinite machines.[30] The problem here is that the infinite individual cannot be repeated or made infinite because that individual is already infinite. Indeed, the self-awareness that attaches to such an individual would merely be in egregious error if it considered itself finite: the fear of harm or death would dissipate if that self-awareness came to realize its infinity. The Fedorovian project could make no sense at all in this context, nor would there be any need for resurrection, since both are based on an erroneous conception of the individual—the Leibnizian individual does not know death but rather a change of orientation in the infinite if it knows any change at all.[31]

Likewise, "infinitization" of the finite human being makes little sense for if Fedorov seeks to assert that the new being is in fact infinite, and not in the Leibnizian way, then we have to ask how such an infinite being can be a being any longer within our understanding of beings as discrete things, that is, finite things having borders such that they are distinct from other things—individuals. An infinite being cannot have such borders because an infinite being has to encompass all. Even if we take the example of the infinite individual in the perhaps most obvious way as a being having an infinity of "connections" with its environment, in its history, etc., we come to the same conclusion because the categorization of these infinite connections must be (by definition) impossible and thus not open to categorization at all. Categorization is of course based on classifications that assume the sharing of a finite set of traits among given things. Clearly, in terms of infinite "things" categorization is unavailable. And, if categorization is unavailable, so is identity, and, if identity is unavailable, there is no identity to preserve or resurrect—Fedorov's project runs aground.

Let us quickly summarize these remarks before proceeding to Fedorov's specific proposals. Two options face Fedorov, one traditional, one radically modern. The immortal being Fedorov seeks to create through technological advancement must be either an infinitely finite being, that is, as a being and as infinite, a repeating set of finite traits or rules, or infinitely infinite such that it cannot be complete or a being

[30] Nachtomy, 113–133.

[31] Here again the charge against the Sophiological approach.

in any conventional sense. Even a modified variant of the infinitely finite being, one that results from the exhaustion of all the possibilities available to that being (such exhaustion being itself predicated on finitude or a sign of finitude) cannot provide another viable option because such a being would not be repeatable. And this of course brings up a serious difficulty with the repeatability option since a repeatable being, as an individual, may in fact not be repeatable at all but only as an example of a species, of a type, of a generality because individuals are, as individual, *not* repeatable.

We have more to say about these difficulties after our presentation of Fedorov's central doctrines, as outlined in his *On the Problem of Brotherhood* and his *Supramoralism* as well as in several of his many essays of various lengths.

Fedorov's clearest political imperative is twofold: (1) the immediate cessation of hostilities among all human beings and the diversion of technological interest to projects that aim exclusively at the bettering of humankind's estate; and (2) the elimination of the distinction between the learned (ученые) and the unlearned (неученые). Both are necessary conditions for creating the common cause and the community that can further the common cause. We examine the basic aspects of this community first before turning to an outline of the common cause itself.

Fedorov refers to this community as a brotherhood, no doubt appealing to the Greek notion of φιλία to express a familial relation among all members of a community. This itself is an interesting choice because Fedorov chooses to create a community that resembles a family rather than the kind of community one finds in Plato or, even in the New Testament where family must be overcome so as to enter into the new Christian community. This thinking is expressed well in Luke and Matthew (specifically Luke 14:26 and Matthew 10:37). Fedorov retains the familial orientation as well as the powerful suggestion that we are all in essence related, that is, we are all united by our identity as human beings such that the notion of individual families separate from all others appears in a harsher light as the result of aggressive (and destructive) sexual competition. In creating his universal family, Fedorov abolishes the product of sexual competition and through immortalization seeks to abolish the sex drive as the origin of human aggressiveness and brutality—in the sex drive lies the model of self-interest and violence to others (a point illustrated powerfully by Sade).

A condition of the creation of this brotherhood is the abolition of the distinction between the learned and the unlearned, in Fedorov's words, between those whose work is purely intellectual or theoretical and those whose work is wholly "physical" or practical. Fedorov seeks nothing less than a sweeping dissolution of the distinction between theory and practice—all work of any kind becomes practical and unified in the pursuit of a single goal, the immortalization of human beings as the step toward the resurrection of all human beings. This dissolution is, one could argue, quintessentially modern, in that theory comes to serve practical ends and gives up its standpoint "above" all practical ends. Ancient θεωρία describes a state of contemplating what does not change, and the theoretical standpoint is one that looks at the world of change and becoming from a static point necessarily beyond it. Nothing could be further from Fedorov's notion of theory, the modern one, where theory provides the basis for promoting practical projects that have as their end the improvement of our estate, the projects that seek, in the clichéd phrase, mastery over nature for the betterment of human life. The same attitude to betterment, the same hostility to nature, are basic components of Fedorov's Christian thinking—Fedorov adapts to what he considers the Christian end of resurrection, the same technological orientation that has often been associated with secularization or the liberation of humanity from Christian "superstition."

Fedorov is thus, at one and the same time, radically modern in this technological sense and, at least in his own self-consideration, thoroughly Christian for the ultimate end is the Christian end of resurrection.

From these two basic principles, brotherhood and the abolition of theory as anything but practical, Fedorov sets up a new universalist community devoted to self-immortalization and the resurrection of all other human beings. We may set this out in steps, as follows, with a few additional measures:

1. Cessation of hostilities among different nations or states leading to the creation of a world state based on brotherhood (and the abolition of divisive sexual desire).
2. Cessation of enmity within the newly created world state by eliminating the distinction between the "learned" and the "unlearned."
3. Transformation of the world state into a full brotherhood unified as such by two tasks: a) the regulation of nature; and b) the preservation of ancestors.

4. Achievement of technological mastery over nature—regulation of nature and empire over the earth, immortalization of human beings.
5. Achievement of perfect recuperation of history in universal resurrection, the filial task of memory being the precondition to universal resurrection.
6. With universal resurrection, the establishment of a final community of deathless *individuals,* full recuperation and abolition of history, empire over death.

Fedorov's view of history is thus distinctive and consistent with his focus on action. History understood as telling stories of the dead, as attempting to preserve their lives in memory, history, in other words, as a form of resurrecting the dead that remains purely *metaphorical* (воскрешение в смысле метафоры) is history for the learned—those who have withdrawn from action, having succumbed to leaving the process of resurrection at metaphor only. History as a recuperation of the past, as a presencing of the past, as a way of cultivating memory, is little more than an admission of defeat in the face of death, little more than an expression of "helpless sadness" because "there is nothing to be done."[32] This kind of history must be suppressed in the common task by a new kind of history that prepares the way for the literal resurrection of the dead by preserving as much as it is possible to preserve of them. History is a *project* that looks forward to the time when universal resurrection will be technically possible—history prepares for this moment and is indispensable to it.[33]

Fedorov's view of history as offering a "metaphorical" resurrection is withering, and we may say the same for those versions of θέωσις that retreat back from literal physical resurrection (likely all notions of θέωσις). The notion of a striving to be deified that never reaches its goal is, evidently for Fedorov, a notion for the learned: the infinite striving towards God, ἐπέκτασις, rather than constituting the way toward deification constitutes, when viewed soberly, the infinite deferral or denial of deification.[34] Fedorov is a revolutionary in his refusal to accept such infinite striving and he is no less withering when it comes to Kant's own

[32] G. W. F. Hegel, *Introduction to the Philosophy of History*, trans. Leo Rauch (Indianapolis, IN: Hackett, 1998), 24.

[33] Fedorov, *What Was Man Created for?*, 77.

[34] Alexis Torrance, 2. Here we add infinite to stretch the point out.

version of infinite striving or Hegel's supposed panlogism in thought that for Fedorov reflects a fantasy based on, and justifying, a failure to act or to understand the role of the learned, of the thinker. In Fedorov's words: "Hegel is the sincerest, the highest and the last philosopher, the thinker par excellence, in truth, the thinker as such. There is nowhere further to go in the area of abstract, all-consuming thinking, idealization of the real. After him, thought should [either end or] transition to action; thinkers should be replaced by men of action, and the 'great' in the realm of thought (in philosophy) will be the 'small' in the realm of action."[35]

Action is crucial. History is a project of destroying death to be realized in action, not thought. History as such is a reflection of filial piety because it preserves the past not as a component of the present or as a way of understanding the present but for the sake of the past in itself. By preserving the past we demonstrate our dedication to preserving our ancestors not necessarily as we see them or as our forerunners—both attitudes that reflect the egoism of a generation that thinks its reality is first and foremost—but as it was for our ancestors. In short, we look back lovingly and grant the past equal dignity with our present or, as we have suggested, we imbue the present with the past, we "presence" or reanimate it. Hence, Fedorov's peculiar fascination with the museum and archive as institutions whose genuine value resides in their reminding us of our duty to bring the past literally back to life, to recuperate the past not merely in memory (though this is itself a necessary beginning) but literally through the discovery of a technology that can return this ostensibly lost life to its original fullness.

The fourth and final part of Fedorov's major text completes the discussion of universal resurrection with an astonishing exhortation to move beyond the confines of our planet by transforming our planet into a sort of space vehicle. By means of this vehicle we may proceed to distribute the resurrected dead throughout the universe, thereby avoiding crowding on this earth.

[35] Fedorov, *Sobranie sochinenii*, 2/109. Our translation.

3 The Power of Fantasy

Is Fedorov's thinking fantastic, outlandish? We should readily admit that it is. And we should readily admit that it is hardly more fantastic than the notion of resurrection that is the core of Fedorov's thought and Christian thought as well (it is Paul, of course, who precedes Fedorov in declaring death as the enemy to be destroyed in 1 Corinthians 15:26). Nor should we fail to admit the profound honesty and directness of Fedorov's refusal to accept the extensive discourse on θέωσις, with its failure to respond effectively to the basic question of what deification of the flesh might literally entail without recourse to a confusing and equivocating plethora of approaches, none of which exceeds "metaphor." Fedorov is indeed the only thinker daring—or foolhardy—enough to demand that the promise of resurrection be met in its literal fullness and not in what he regarded as mendacious intellectual subterfuges that permitted the learned to escape the disappointing reality of their failure to insist on transformative action, a failure that only abetted the continued flourishing of the kingdom of darkness and death.

Resurrection is the cornerstone of Fedorov's thought and Fedorov's adamant insistence on literal, physical resurrection, a restoration to life of the death that has caused so much travail and despair, is bracing—it forces his theological opponents into a theological corner. For they must claim either that resurrection is unfeasible or undesirable in its literal form. Fedorov's opponents, in other words, are forced into a defense of death based on an assumption that resurrection is a mystery of some kind, a metaphor involving various "theories" or "narratives" whose task it is to explain why literal resurrection is to be dismissed. And it forces everyone else into the Schmittian corner of favoring the continuation of history in all its violence and horror.[36]

The potential weakness of Fedorov's view lies, as he realized, in an accusation that it is an expression of egoism, or worse, of a covert atheism that seeks to obliterate the distance between human beings and God, transforming the created being into one equal to the uncreated being. This accusation against Fedorov understands θέωσις as not entailing that

[36] On Carl Schmitt's *katechontic* understanding of history, see Julia Hell, *The Conquest of Ruins: The Third Reich and the Fall of Rome* (Chicago: The University of Chicago Press, 2019), 403–430.

the human being literally becomes God but that they give their self to God in a kenotic act whereby the human being becomes free of the selfish inclination that turns the human being away from God and prevents the transformation of the flesh from mortal existence to immortal existence. The evident objection to this understanding of θέωσις is that, while it might be attractive insofar as it emphasizes a transformation of orientation in the human being, it remains within the precincts of metaphor or of infinite striving. The basic problem of infinite striving is that it is infinite precisely because selfish inclination cannot be overcome, at least not in a being that continues to nurture the flesh in order to survive.

Survival as a barrier to true communion with God explains the association of deification with a death drive by those who critique Fedorov not only for his egoism but also for his nihilism. Rather than becoming immortal, so this critique holds, deification involves a radical impoverishment, if not annihilation, of human life in its singular attempt to become like God—to be like God is in fact to reject one's humanity to the extent the transformation of the flesh crucial to deification leads human beings away from the simple effort to survive that requires violence and destruction of other beings, no matter how mild those may seem to be. Alexandre Kojève offers a potent critique of Fedorov's position, and indeed of deification in general, when he indicates that the human can be neither fully immortalized nor severed from its finite "substrate":

> The human being, therefore, is just as *real* and *objective* as the natural being. And he is just as *actual* (or active) as the latter. One cannot say, however, that there is a dualism here in the ordinary sense of the word: man is not a *substance* opposed to natural substance. For man is only the *negation* of Nature: he is *Negativity*, and not *Identity* or Substantiality. If the Universe is a ring, and if Nature *is* like the metal of which this ring is made, Man *is* only like the hole of this ring. For the Universe to be a ring and not something else, there must be a hole in the same way as the metal which surrounds it. But the metal can exist without the hole, and the Universe would be something without being a ring. The hole, by contrast, would be nothing without the metal, and there is no Universe that would only be a hole. Thus, Nature can exist without Man, and a purely natural Universe is perfectly conceivable (although its conception in actuality presupposes in fact the existence of man who conceives it and

> who this Universe includes). A purely human Universe, by contrast, is inconceivable, for without Nature, Man is nothingness pure and simple.[37]

The human being in this sense is pure activity—to make the human being immortal, a static being, would be to extract its humanity from it, to transform the human being from a being that is radically free insofar as it negates as opposed to a being that is purely static and, thus, purely finite and unfree. What Kojève points out is the essential inconcinnity of immortality and finitude: the static being can only be bereft of consciousness of itself (which would rob it of its stasis). To render a human being immortal must in some fashion be to conserve that being in its entirety and, as such, as a kind of static being that can no longer be human or recognizably free but rather finds itself (and would it find itself if completely objectified?) utterly trapped within itself in a more horrendous trap than our finite lives to the extent that trap is unending. Immortality, viewed from this perspective, is predicated upon the complete objectification of the human being and thus, in Kojève's terms, represents the destruction of its humanity.

Kojève's comments have the virtue of making one consequence of Fedorov's brotherhood clear: that it must be utterly static. Fedorov's insistence on resurrection is in fact apocalyptic in the sense that it seeks to restore all of history in a *fixed and immutable present*. To speak of this present as if it in any way could resemble our everyday existence in time, our living being, would be to falsify the implications of Fedorov's brotherhood and ignore its apocalyptic hope of ending time (ὅτι χρόνος οὐκέτι ἔσται).[38] The problem is that human being as such exists in time, not outside it, and to imagine a human being no longer subject to time but still in time involves a contradiction and imposes a serious difficulty on the imagination that has to project a human being that is similar to the living being in time as a being freed from time. And the freedom from time seems to rob the human being of what is human, if we equate human being with freedom and negativity. If we do not, we impose an identity on the human being that takes away all freedom, individuality and likely consciousness as well: the imposition of immortality ends up imposing something akin to animal or machine being that is neither free nor truly dynamic but a being defined by stasis and repetition—such a

[37] Alexandre Kojève, *Outline of a Phenomenology of Right*, trans. Bryan-Paul Frost and Robert Howse (Lanham, MD: Rowman & Littlefield, 2008), 214.

[38] Revelations 10:6.

life, though eternal, would seem to resemble the Tartarus of the Greeks or the hell of Dante whose denizens are all trapped in a repetition of the earthly lives of their sins, which as natural desires, tend to repetition or compulsion.

Kojève highlights as well an important notion: the negative. While Soloviev rejects the negative in favor of his positive absolute, it is not inappropriate to suggest that a critique of Soloviev's positive absolute emerges in Kojève's comments in the contradictory currents apparent in the notion of infinite striving for a deification that never takes place—the essence of infinite striving is the very negativity that Soloviev decries and Kojève identifies as the crucial distinguishing aspect of the human, which strives to negate its negativity or emptiness. Human striving never ends in a final, positive identity because no human is ever finished creating positive content for itself. There is always more to be done, always more possibilities of negation so long as one is alive:

> [Man's] possibilities go beyond all his actual realizations and are not determined by these realizations in an unequivocal manner. But this is also to say that he can actually realize only a limited number of his infinite (or better: indefinite, in the sense that every non-A is indefinite) possibilities. In other words, man always dies somehow prematurely (which to a certain extent "justifies" his desire for an afterlife) — that is, before exhausting all the possibilities of his being (or better: of his negating or creative action).[39]

Though the possibilities of human negation might be infinite, for individuals time is never infinite, as everyone knows: everyone knows and feels the "rush" of time. But this negative intrusion of the finite into the infinite constitutes the essence of human life. Without death, there would be no action, no history, for human life consists in striving to negate one's own future annihilation. Human life consists in striving to free oneself from death as "a manifestation of man's negativity," and striving to free oneself from death only manifests itself negatively through the individual human actions of history (*Geschichte*).[40] In short, human freedom is historical and alive, not static and dead, because human life is

[39] Alexandre Kojève, *Introduction to the Reading of Hegel*, trans. James H. Nichols, Jr. (Ithaca, NY: Cornell University Press, 1980), 256–257.

[40] Ibid., 254.

temporal and dynamic.[41] To make human life eternal is to negate it, to extinguish it.

The negative, if the true identifying marker of the human, is thus crucial. Heidegger is perhaps the most formidable modern advocate of the negative and his destruction of the metaphysical tradition bears a close resemblance to Kojève's notion of negativity that is itself likely inspired by Heidegger. We turn to Heidegger's treatment of the negative now both as a foil to the project of deification and as an anticipation of our discussion of the problem of technology in China.

Heidegger offers what is perhaps the most cogent and wide-ranging critique of the drive to mastery of nature, to "regulation" of nature in Fedorov's terms, that places emphasis on negation as crucial to our world-understanding. The significance of this focus on negation comes clear when we consider that immortality absolutely requires pure presence, the very presence that constitutes the most powerful heritage of Greek thought for Heidegger and provides a stark contrast to the "Eastern" privileging of the negative. Such a pure presence is not in fact the highest expression of human life, but its negation—for who has ever lived outside of time? Fedorov marks a radicality matched by Heidegger's equally radical insistence on finitude, both filtered through a tension in modern Chinese thinking that may indeed be immensely productive. But we anticipate. The unusual question we may pose here as a prefatory comment to the next chapter is: Why should we reject Fedorov's common cause? What are the underlying reasons for Heidegger's alarming account of modernity? What is "wrong" with an anthropomorphic transformation of nature? Why should we not exploit nature as much as we can for our benefit? Are these questions, as we have already noted, of feasibility or of morality, of what we should and should not do? Fedorov's moral imperative seems clear. Can we say the same of Heidegger's?

[41] Kojève says as much himself when he notes that suicide "'manifests' freedom, it does not *realize* freedom, for it ends in nothingness and not in a free *existence*" (ibid., 248; see also Alexandre Kojève, *Atheism*, trans. Jeff Love (New York: Columbia University Press, 2018), 155 [note 136] and the quote cited and translated from the French edition of Kojève's Hegel lectures in note 36 of chapter three of this book). As we discuss in that chapter, Heidegger suggests a similar view of the human with his distinction between *Geschichte* as the vital, creative energy of *Dasein* and *Historie* as the dead remainder of that energy.

CHAPTER 3

Recuperation of the Finite

Martin Heidegger (1889–1976) seeks to think what cannot be thought. This approach is nowhere clearer than in his various attempts to think the nothing (*das Nichts*) or think what is the contrary of any being and beings as a whole. How can it be possible to think what is not a being, what is not "there" at all, so to speak? The implications of this manner of thinking are momentous because it resists both our most commonsense attitudes and, more basically and incisively, the attitude that orients us to sense rather than to nonsense or what is "other" than sense. To orient us away from sense sounds radical indeed, and it is, if sense becomes the definitive measure of our attitudes to what we encounter in the world. For if the measure of a being is that it makes sense to us, then that being is only a being to the extent it makes sense. What fails to make sense gets relegated to nonsense, madness or speech that is equivalent to silence, a silence that in its refusal to open up to sense is indistinguishable from what is nothing for us. Nothing does not speak to us, nothing does not make sense, nothing eludes the nets of reason and, as such, it eludes our attitudes to beings in the world if these are in fact fundamentally cognitive.

Why think the nothing? Why does Heidegger take such an approach to thinking? These are among the guiding questions of this chapter. We give an account of this most radical aspect of Heidegger's thought in order to provide the necessary context for understanding his distinction

J. Love and M. Meng, *Revolutionary Bio-politics from Fedorov to Mao*, https://doi.org/10.1007/978-981-99-4745-4_3

between Historie and Geschichte, along with its wider significance for the broad narrative of deification we have introduced in the preceding chapter, especially in terms of Soloviev's rejection of the negative absolute.

We begin by discussing Heidegger's account of nothing anchored in "What Is Metaphysics?" (1929). We then proceed to Heidegger's related treatment of transcendence and truth by reference to "On the Essence of Ground" (1929), "On the Essence of Truth," (1930/1943) and *Contributions to Philosophy* (1936–1938). Finally, we develop the distinction between Historie and Geschichte in detail, starting from the *Contributions to Philosophy*. Our discussion is shaped by an intention to show how Heidegger emphasizes finitude in opposition to the "immortalizing" currents we discussed in the preceding chapter as both opposition to those currents and transition to the final chapter on Chinese modernity.

1 The Nothing

Heidegger's unusual discussion of the nothing in "What Is Metaphysics?" elicited a striking response from Rudolf Carnap, among others.[1] Carnap's response elucidates some of the fundamental issues under consideration in this section of the book. Accordingly, we want to frame our discussion of Heidegger by reference to this response, which, in our view, brings out most clearly and consequentially what is at stake in Heidegger's attitude.

"What Is Metaphysics?" is an enquiry into the nothing that affirms the basic importance of the nothing to our everyday relation to beings or things in the world. Yet, Heidegger begins by indicating that this relation tends to be ignored, especially in our cognitive attitude to beings. "Science wants to know nothing about the Nothing."[2] Science cannot banish the nothing, however, since it is implied in the very notion of a being as what is and thus what is *not* nothing. The most simple declaration of the being of any given thing presupposes that the being is something and not nothing. As a result, the nothing accompanies any and every aspect of a given being, and it does so not only in terms of the being's existence but also in terms of the various predicates that

[1] Rudolf Carnap, "Überwindung der Metaphysik durch logische Analyse der Sprache," *Erkenntnis* 2 (1931): 219–241.

[2] Martin Heidegger, "Was ist Metaphysik?" in *Wegmarken* (Frankfurt am Main: Vittorio Klostermann, 1976), 108. Unless otherwise indicated, all translations from Heidegger's German are our own.

apply to the being. This is the case because the predicates depend on the being's existence. Hence, the typical distinction in the copula "is" between an existential and a predicative use becomes complicated since the one depends on the either.

This distinction is important for understanding the nothing because we may distinguish between the nothing presupposed in the existential use of the copula and that which may be deployed in predicative use. In the former case, the nothing, as what is other than the being itself, is absolute in the sense that it is wholly other to the being as "pure" non-being or nothing: it is the *nihil absolutum*. In the latter case, the nothing, or "not," may be deployed to differentiate the being from other beings, that X is not Y because it is different from Y at least in regard to one predicate, though of course other predicates may apply to express that difference as well. This is then the *nihil relativum* which indicates relative difference, of one being from another, rather than the absolute distinction drawn by the difference between a being's existence as such and its non-existence. That X is indicates an absolute distinction from X's non-being—the nothing that X is not—and not a relative difference between two beings. Put most succinctly, there is a radical distinction to be made between X and nothing as nothing and X and Y as different.

For Heidegger the dependence of predication itself on the existence of a being is reflected in the structure of negation as well, for the negation that we have referred to as relative, the negation pertaining to predicative difference, is dependent on the existence of the being to which it applies. Accordingly, this negation depends on the more radical form of negation that does not say that X has a predicate that does not apply to Y but rather that X has no predicates at all because it does not exist. Here we enter into the greater problematic of Heidegger's essay and Carnap's critical comments as well. The problem may be stated simply: How can one say that X is not—absolutely—without contradicting oneself? Have we not indicated in this absolute negation that the object of the negation, X, in some way is? How can one ever say that something is not without attributing being to it and thus cancelling the negation by contradicting it?

Heidegger is much more provocative. He asks: "How does it stand with the nothing?"[3] The question perplexes. Heidegger does not ask,

[3] Ibid., 106.

"What is the nothing?" but rather "How does it stand with the nothing" ("Wie steht es um das Nichts?"). The first question points clearly to the problem: if the nothing "is" nothing, how can the interrogative "What?" apply to it? "What" expects a definition, and a definition of something requires predicates. If the nothing is defined as what has no predicates—and how else could we understand the nothing?—then the question quite obviously cannot be answered other than negatively in the sense that nothing is not a "what." Hence, Heidegger takes a different approach by asking *how* it stands with the nothing. Use of the verb "to stand" complicates our understanding. Why does Heidegger not simply ask: "How is the nothing" or, as David Farrell Krell translates it: "How is it with the nothing?"[4] Heidegger's question is typically ambiguous, employing a phrase "How does it stand with X?" both in its ordinary meaning in German as an enquiry into the state of a matter but also in its other more literal meaning as "How does it stand around the nothing?" The German preposition "um" is a significant one for Heidegger; and if we interpret more literally, we may consider the sentence as posing a question about how things stand, that is come forth or arise, around the nothing, the absent center or origin. One may object that such a reading is fanciful since standing is merely a state and "around" in this sense has to become metaphorical insofar as it describes a spatial relation to what cannot be in space. And perhaps "fanciful" is too generous a term to use here: Carnap takes the objection much farther, though he does so in connection with an even more provocative phrase of Heidegger's that appears in his response to this question.

The phrase, "Das Nichts selbst nichtet," is more difficult to translate: "The Nothing itself noths" (Inwood), "The Nothing itself nihilates" (Krell), or, indeed, "The nothing nothings."[5] By means of this construction Heidegger offends both against the "what-question" and the "how-question." What nothings? The nothing. How is it with the nothing? It is by nothing (nothinging) or "nihilating."[6] In the first case,

[4] "What Is Metaphysics?" in *Pathmarks* ed. William McNeill (Cambridge: Cambridge University Press, 1998), 82–96.

[5] Michael Inwood, "Does the Nothing Noth?" *Royal Institute of Philosophy Supplement* 44 (March 1999): 271–290; Krell's translation is in *Pathmarks*, 82–96; and "the nothing nothings" is our preferred translation.

[6] The participial construction reveals why translation is so vexed. Krell's awkward choice avoids the absurdity of the participle of "noth" ("nothing") or "nothing" ("nothinging").

Heidegger accentuates the difficulty, not by saying that "X is not" in the existential sense, but by saying that Nothing is, and he capitalizes the word in German to underline its status as a noun. Moreover, rather than saying that the "Nothing is," he utilizes a neologism in German to suggest that the noun also describes an activity. Carnap suggests that this structure is not so unusual, one may say "the rain rains" or the "question questions."[7] But other examples bring out the radical strangeness of the phrase more emphatically: "the dog dogs," "the book books," and the like. These sentences show that Heidegger challenges grammatical categories, specifically of noun and verb.

For his part, Carnap prefers to address the logical structure of the phrase. His basic argument is that the phrase, while appearing in the guise of sense, is in fact nonsense. It is nonsense because the noun cannot refer to any being nor can the verb refer to any activity. Carnap's assumption is that "nothing" cannot act as a noun because it names no determinate referent. "Nothing" is no clearer than the word "God"—one cannot point to nothing, nor can one point to God, as defined, determinate beings or objects of experience. Hence, "nothing" is not like "rain" or "dog" insofar as one cannot refer to it as a specific being or category of specifically identifiable beings, like weather events or animals. Since "nothing" exceeds the bounds of all possible experience, at least as cognitively available to us, it cannot be a logically valid subject in the sentence. By the same logic, and without regard to its grammatical structure, "nothing" cannot be transformed into an activity, since, like the noun, the activity has neither determinate content nor referent. Indeed, the act of "nihilating" could be understood only negatively in terms of "an-nihilating" or destroying a determinate being or set of beings. The negation in that case takes form, so to speak, only negatively in the destruction of a given form. Whether such a negative process can have any positive identity depends of course on another assumption as to whether destruction is absolute, a turning of something into nothing, or only relative, *transformation* of a thing into something else.

Carnap holds then that "nothing" cannot be meaningful within cognitive terms. He does admit that "nothing" as *affective* may have some meaning, but that meaning does not belong within the purview of theory or science as cognitive activities but within poetry or music. Carnap

[7] Carnap, "Überwindung der Metaphysik," 230.

famously derides metaphysicians as "musicians without musical ability" who clothe in the language of scientific discourse assertions that do not comply with the standards of coherence or meaningfulness of such discourse.[8]

Carnap's critique is not easily dismissed, nor should it be. While he is quite aware that Heidegger recognizes the extent of his provocation, Carnap nonetheless refuses to accept the critique that lies within that provocation. The basic contention concerns logic itself. Carnap asserts the validity of logic and defends logic against Heidegger's implicit critique. This critique consists in his claim that logic is simply unable to comprehend a sentence like "The nothing nothings." Logic is an interpretation of language and thinking that imposes restrictions on both and excludes usage which does not conform to its rules. If, for Carnap, logic is precisely this assertion of rules, an attempt to liberate language from obscurity and confusion, for Heidegger logic reduces and imprisons—it takes for granted what we dare not take for granted, and nowhere does logic do so more flagrantly than with the nothing. That logic must exclude the "nothing" as nonsense, rather than attempt to think it, is the weakness of logic which cannot make a claim to know or rule the whole if it cannot think the whole itself transparently but only beings *within* the whole.

Now, it may seem that we have made a precipitous transition from logic as failing to understand the nothing to logic as failing to understand the whole. But this transition brings us to the full import of Heidegger's provocation and critique of logic. For if logic can think only in terms of beings, it cannot think the horizon within which beings become accessible to us as such. Logic can categorize, delineate and define beings and their relations to each other, but it cannot explain or even refer to that within which such beings appear to us unless the latter could be defined as a being as well. This "container" or "set" of all beings is the most problematic being of all, and Heidegger wants to bring that problem out into the open, a problem which he claims logic does not clarify or even take into account since logic, in accordance with its own governing rules, is totally unable to do so. The basic point is this: that which contains all beings is not itself a being. If it were, it would not contain all beings but would in fact presuppose yet another being in which it is contained, thus moving the final being back another step *ad infinitum*. In other words,

[8] Ibid., 240.

if every being implies its own non-existence or a more extensive non-existence when applied to all beings, then how can one ever come to a final positive description of the whole, a fixed totality like that implied by Fedorov's project of resurrection? Any such description would have to presuppose a negation that it cannot by definition include. Moreover, the mere possibility of thinking all beings or a final picture of beings, undermines that finality and reveals that there is a basic division between beings as they are and the being that perceives them as such—this being has to transcend beings in order to grasp them; otherwise, there would be no way to grasp beings as beings—beings in the final world order cannot be conscious of themselves as such.

We want to clarify this point because it is of utmost significance. Why does the grasping of beings as beings require transcendence, awareness, or transcendent awareness? For one thing, the mere perception of a being as such suggests there is a context in which the being appears that "exceeds" it. We perceive another being within a context that relates us both and is not reducible to that relation. Why must this be so? Can we not perceive another being in itself regardless of context? Can there not be an immanent plane that lines beings up and relates them to each other? How are we to know that we perceive and other beings do not perceive us? Evidently, we cannot transcend their difference from us so as to know that they see us or other beings from their perspective. Hence, we cannot make a determinative conclusion that they do or do not perceive us. But we can conclude that they are there before us and that we interact with them in various ways. These interactions themselves may constitute the context in which we encounter each other, and to know things as a whole we do not need recourse to negation but rather to a process of addition, perhaps—but this is unprovable—*infinite* addition, as it is exactly the beings we encounter which make up the whole. Being is thus the sum of all beings, not some mysterious and nonsensical nothing we can neither think nor perceive.

Heidegger does not accept these arguments. The whole is not a sum of its constituent parts nor the totality of beings, at least for the being that we are. The basic argument follows the lines we briefly sketched out above. If one simply adds beings to reach a final sum, one comes to something like the set of all sets, and, as we know, the set of all sets, the largest set, brings with it insoluble problems. The so-called Russell paradox puts the matter succinctly. The set of all sets, as the largest set, cannot belong to any set other than itself. But the largest set must and cannot be a member of itself. As Russell puts it:

> The comprehensive class we are considering, which is to embrace everything, must embrace itself as one of its members. In other words, if there is such a thing as "everything," then, "everything" is something, and is a member of the class "everything." But normally a class is not a member of itself. Mankind, for example, is not a man. Form now the assemblage of all classes which are not members of themselves. This is a class: is it a member of itself or not? If it is, it is one of those classes that are not members of themselves, i.e., it is not a member of itself. If it is not, it is not one of those classes that are not members of themselves, i.e. it is a member of itself. Thus of the two hypotheses – that it is, and that it is not, a member of itself – each implies its contradictory. This is a contradiction.[9]

Even without classes, if we merely enumerated beings without regard to their specific identities, we would encounter similar difficulties since determination of the final being suggests itself a position or place beyond that final being that cannot be added into it—the whole is not merely a sum or biggest number if these are both understood as an agglomeration of beings. There is no highest number. The whole of beings cannot itself be an enumerated being and be the whole of beings—the being of beings is not itself a being.

Here we have two closely related arguments that end in contradiction. The biggest set cannot be a set within the definitions applicable to sets just as a highest number cannot be a number, either a finite or infinite one (if such were possible)—as Leibniz notes in the latter case:

> If the numbers are assumed to exceed each other continuously by one, the number of such finite numbers cannot be infinite, for in that case the number of numbers is equal to the greatest number, which is assumed to be finite. It has to be replied that there is no greatest number. But even if they were to increase in some way other than by ones, yet if they always increase by finite differences, it is necessary that the number of all numbers always has a finite ratio to the last number; further, the last number will always be greater than the number of all numbers. From which it follows

[9] Bertrand Russell, *Introduction to Mathematical Philosophy* (New York: Dover Publications, 1993), 136. Russell's paradox applies explicitly to a logic of classes; in terms of sets, Georg Cantor seems to have come to a similar conclusion several years prior to Russell.

> that the number of all numbers is not infinite; neither, therefore, is the number of units. Therefore there is no infinite number, or, such a number is not possible.[10]

The whole as an enumerative totality or as a largest class—a container holding everything or everything itself—produces contradictions that reveal the limits of largely mathematical logic, the same logic that comes to dominate the investigations of Carnap and those influenced by him. Heidegger's response to Carnap is clear, and aggressive. In the *Introduction to Metaphysics* from 1935, Heidegger offers a pointed dismissal without naming Carnap (at least in the published text)[11]:

> One cannot, in fact, talk about and deal with Nothing as if it were a thing, such as the rain out there, or a mountain, or any object at all; Nothing remains in principle inaccessible to all science. Whoever truly wants to talk of Nothing must necessarily become unscientific. But this is a great misfortune only if one believes that scientific thinking alone is the authentic, rigorous thinking, that it alone can and must be made the measure even of philosophical thinking. The reverse is the case. All scientific thinking is just a derivative and rigidified form of philosophical thinking. Philosophy never arises from or through science. Philosophy can never belong to the same order as the sciences. It belongs to a higher order, and not just "logically," as it were, or in a table of the system of sciences. Philosophy stands in a completely different domain and rank of spiritual Dasein. Only poetry is of the same order as philosophical thinking, although thinking and poetry are not identical. Talking about Nothing remains forever an abomination and an absurdity for science. But aside from the philosopher, the poet can also talk about Nothing—and not because the procedure of poetry, in the opinion of everyday understanding, is less rigorous, but because, in comparison to all mere science, an essential superiority of the spirit holds sway in poetry (only genuine and great poetry is meant).[12]

[10] G. W. Leibniz, *Philosophische Schriften* AAVI, 3 (Berlin: Akademie Verlag, 2006), 477.

[11] The edition of the *Introduction to Metaphysics* published as volume 40 of Heidegger's *Collected Edition* (*Gesamtausgabe*) includes a second appendix in which Heidegger responds by directly naming Carnap.

[12] Martin Heidegger, *Introduction to Metaphysics*, trans. Gregory Fried and Richard Polt, 2nd ed. (New Haven, CT: Yale University Press, 2014), 28–29.

Logic and science cannot speak adequately about "Nothing"—this we have seen. But the connection of "Nothing" with the whole, which can be nothing for logic and science, takes us further into Heidegger's primary investigation into our relation to beings, the relation that presupposes a context that transcends that relation. Heidegger seems to accept the problems of an enumerative whole as merely taking all things together as they are, both in terms of Russell's paradox and Leibniz's notion of a highest or final number. These paradoxes simply show that logic (or science) is indeed unable to think the whole as somehow beyond being. Moreover, they are unable, as such, to think the whole as the context in which we encounter beings or first relate to them—this context, derided as nothing or nonsense in terms of logic or scientific thinking—is for Heidegger the crucial matter for thought that eludes thought to the degree it is confined to logic. Hence, Heidegger's reference to poetry, a reference that stymies some commentators while leading others to accuse Heidegger of being an irrationalist or mystic. Such accusations, however, are based on faith in the very logic that Heidegger seeks to put in question or transcend. This peculiar word, with so many associations, becomes important for Heidegger in terms of how he reconfigures "Nothing" as one term for the context in which we come to relate to beings.

2 Transcendence

This relation is what Heidegger refers to as transcendence in the 1920s, a relation he complicates in his later work. The relation transcends beings and the logic of beings: it points to what Heidegger refers to in "On the Essence of Truth" as the open (*das Offene*) or freedom. For Heidegger associates transcendence with freedom and with fundamental, not propositional truth, as freedom. His striking claim is that "the essence of truth reveals itself as freedom" ("*Das Wesen der Wahrheit enthüllt sich als Freiheit*").[13] Heidegger articulates the connection between transcendence, truth and freedom in a number of writings in the vast corpus of his published and unpublished work. For our purposes, it is sufficient to examine two central and concise texts from what some scholars consider his "transitional" period (perhaps 1927–1932): "On the Essence

[13] Martin Heidegger, "Vom Wesen der Wahrheit," in *Wegmarken* (Frankfurt am Main: Vittorio Klostermann, 1976), 192.

of Ground" (1929) and "On the Essence of Truth" (1930 but published first in substantially revised form in 1943).[14]

Still, to orient this examination, Heidegger's programmatic statement about transcendence from *Being and Time* (1927) is useful:

> Being, as the basic theme of philosophy, is no class or genus of entities; yet it pertains to every entity. Its 'universality' is to be sought higher up. Being and the structure of Being lie beyond every entity and every possible character which an entity may possess. *Being is the transcendens pure and simple.* And the transcendence of Dasein's Being is distinctive in that it implies the possibility and the necessity of the most radical *individuation*. Every disclosure of Being as the *transcendens* is *transcendental* knowledge. *Phenomenological truth (the disclosedness of Being) is veritas transcendentalis.*[15]

Unlike Kant or Husserl, Heidegger displaces transcendence: far from pertaining primarily to the subject and subjective consciousness, transcendence pertains to Being itself as what is not a being, thus no-thing or the open, the presumptive "context" in which beings are disclosed that subtracts itself from them precisely *as* disclosed. Nonetheless, Heidegger also refers transcendence to Dasein, thereby preserving some element of the transcendental subjectivity of Kant and Husserl. It is this aspect of transcendence that proves most problematic for Heidegger himself because it raises doubts about whether Heidegger can wrest transcendence from the Kantian and Husserlian tradition, thereby avoiding a return to the subject or to a broadly anthropological reading to which Heidegger remains steadfastly and explicitly opposed.

In "On the Essence of Ground" Heidegger identifies transcendence with "surpassing" (*Überstieg*) whereby Dasein moves beyond itself to beings. Dasein encounters these beings, individually and as a whole, in a broader "context" that Heidegger calls world. Heidegger is thus

[14] "On the Essence of Truth" was originally presented in the form of a talk that Heidegger gave on four occasions in 1930. He revised the talk subsequently at various points in the 1930s and completed a substantial revision in 1940, though he did not publish the final version until 1943. The signal importance of the talk for Heidegger is indicated by his frequent reference to it in other works. William Richardson considers the talk important enough to signal the beginning of a new period in Heidegger's thought.

[15] Martin Heidegger, *Being and Time*, trans. John Macquarrie and Edward Robinson (Oxford: Blackwell Publishers, 1962), 62.

careful to differentiate the whole, as the whole of things, from the world. He does so because world surpasses things, as it must if they are to appear within it. Likewise, world must also surpass Dasein insofar as Dasein, too, appears in the world. Dasein, as such, is unusual in that it is not only surpassed by the world but does itself surpass beings. In this respect, Heidegger indicates that surpassing—or transcendence—applies both to world and Dasein. The problem is how to work out the relation of Dasein to world as a transcendental one, and Heidegger seeks to avoid an interpretation which confers on Dasein primary transcendental responsibility. If Dasein were world disclosing, then Dasein would seem to pre-exist the world. But Heidegger assures us that this is not the case: Dasein and world are mutually disclosing, the one cannot function without the other. In other words, Dasein presupposes world, and world presupposes Dasein, and both presuppose transcendence. What then "is" transcendence?

Here is the problem: if transcendence surpasses beings and the definition of being inheres only in beings, there is no way to define the being of transcendence in terms of beings. Yet, if there is no way to define the being of transcendence in terms of beings, then transcendence would seem to invite the same opacity as the notion of "nothing," the world, or the "whole" as world, i.e. as that which surpasses world. Heidegger does not want to identify the transcendence applicable to Dasein or world with God, the Medieval definition, or with a subject whose conditions of access to the world precede it a priori or otherwise condition it so that it is these conditions of access that define what is as the basic conditions for its appearance as such.[16] This model of subjective transcendence is inferential in the sense that one has to infer conditions for the appearance of beings that do not themselves appear as beings or, at least, in the same manner as beings. For Heidegger these definitions of transcendence lead to insuperable difficulties because they presuppose a constituting agency that does not appear directly in what it constitutes. But that is not the only difficulty: the attribution of constitutive agency to a human subject comes perilously close to a claim that human beings in fact create the world or that the world is a mere epiphenomenon of the creative subject

[16] Martin Heidegger, *Logic: The Question of Truth*, trans. Thomas Sheehan (Bloomington: Indiana University Press, 2016), 90–103; Martin Heidegger, *The Metaphysical Foundations of Logic*, trans. Michael Heim (Bloomington: Indiana University Press, 1984), 43–50.

rather than crucial to its constitution. And it may be fair to say, in rough terms, that Heidegger's ambiguous attitude to transcendence, at least in part, results from hesitancy to grant creative priority either to the subject or the world, usually understood as nature in the role of transcendent constitutive agency.

In "On the Essence of Ground" Heidegger is thus compelled to describe almost a "third" space, beyond both Dasein and world, assuring their capacity for transcendence. He associates this "third" space with freedom and possibility. Yet, it is Dasein to which possibility is relevant. To avoid ascribing to Dasein creative agency, however, Heidegger maintains that the world itself is the "space" of possibility to which Dasein relates as such. Transcendence as freedom is possibility, not possibility as relating to one being, but as completely open possibility—akin to nothing in its indeterminateness. Dasein is transcendent because it relates to this possibility, and the world too since it is the realm of possibility which itself lies beyond any given determinate possibility (such as that an acorn may or may not become an oak tree):

> All forms of comportment are rooted in transcendence. The "will" in question, however, must first "form" the "for the-sake-of" itself as and in a surpassing. Yet whatever, in accordance with its essence, casts something like the 'for the sake of' projectively before it, rather than simply producing it as an occasional and additional accomplishment, is that which we call freedom. Surpassing in the direction of world is freedom itself. Accordingly, transcendence does not merely come upon the 'for the sake of' as anything like a value or end that would be present at hand in itself; rather, freedom holds the 'for the sake of' toward itself, and does so as freedom. In this transcending that holds the 'for the sake of' toward itself there occurs the Dasein in human beings, such that in the essence of their existence they can be obligated to themselves, i.e., be free selves. In this, however, freedom simultaneously unveils itself as making possible something binding, indeed obligation in general. *Freedom alone can let a world prevail and let it world for Dasein*. World never *is*, but *worlds*.[17]

Freedom that is indeterminate, just like indeterminate possibility, brings us back to the concerns expressed by Carnap as to whether such freedom or possibility is anything other than nonsense. If sense is defined by determinacy relative to beings, then it is quite clear that anything that

[17] Heidegger, "On the Essence of Ground," 126.

is indeterminate, thus not a thing as such, can be only contradictory or nonsensical. Does the same apply to freedom?

As we noted above, "On the Essence of Truth" declares that "truth reveals itself as freedom." One may well accuse Heidegger of provocation once again: How can truth be freedom? What sort of strange combination can this be? Heidegger raises two challenges with this sentence. On the one hand, he ties truth to freedom and freedom, in the final account, to indeterminacy, the realm of "pure" possibility. What could be more estranging than to argue that truth, as such, is indeterminacy? Is this not equivalent to saying that the truth is nothing, or that the truth nothings?

Let us examine the essay on truth in a bit more detail prior to responding to this question. Heidegger opposes two different notions of truth, both as responses to the simple question: What is truth? The first response advances the notion of truth as based in a relation of correspondence between a statement or proposition and a thing. If we point to a coin on a table and say "that it is a coin," and it is indeed a coin, then we have made a true statement because the general term "coin" denotes or refers to the specific thing we see lying on the table. The truth of my statement arises from the relation of correspondence between my statement and the thing: we have correctly matched statement and thing. Hence, Heidegger refers to this relation as that of correctness or even equivalence (*adaequatio*), the latter being the basis for the former. Correctness, Heidegger notes, presupposes another notion of truth because the thing to which the statement refers must first be disclosed as the thing it is for the statement to be correct. There can be no correctness without this prior disclosure that sets the norm required for correctness. Heidegger claims, however, that truth as correctness tends to conceal truth as disclosure. The normative aspect of truth assumes an encounter that is normative insofar as it creates the basis for a norm, but also non-normative in that there is no way to judge of the correctness of the disclosure itself: the norm is grounded in a disclosure that is not itself subject to that norm or any other for that matter. The table as we understand or picture it is not necessarily a correct disclosure because we do not know what a correct disclosure would be. The somewhat shocking conclusion to be drawn here is that normativity arises from a non-normative disclosure in the sense that the latter is not subject to a set of rules but is in effect "free."

And this assertion brings us back to the beginning of our discussion and the thing itself as disclosed. Heidegger cautions that lack of rules, freedom, is not to be equated with the merely arbitrary. His caution in turn seems to have to do with the same concerns about the subjective that Heidegger expresses in regard to Dasein's transcendence insofar as Dasein cannot simply arbitrarily create the thing out of thin air or its own immanent consciousness. To the contrary, the mutually determinative relation that makes out and depends on world is determinative as that within which any disclosive encounter may take place and, as such, it cannot be subject to determination as the ground of determination (this would set up a *reductio ad infinitum* precisely cognate to that between the normative being and its origin). Dasein is unable to create the world as God does, and transcendence as we have viewed it, requires this fundamentally prior indeterminacy as the basis for any determinacy that Dasein works out.

This indeterminacy is the freedom of truth. It cannot be any "thing" because it is the basis of things that both appears in and transcends them as such. It is indeed easy to read Heidegger as perpetuating the arbitrary by suggesting that things emerge without any reason *as* they do: we cannot say why a tree is a tree and not a star any more than we can explain—other than through causal mechanics why a rose blooms. The temptation is of course to develop narratives or, alternatively, mechanical accounts that confer an explanatory teleology on beings. The effort we make, according to Heidegger, is to cover over the explanatory abyss from which beings come. This terminology is hyperbolic, to be sure, but it does point to the guiding utility of explanatory schemes that seek maximally to suppress the absence of reasons for the disclosure taking the exact form it does.

Heidegger's account is opposed to those, such as Kant's, that apparently locate the normative basis for things existing as they are in the a priori conditions at work in the subject's cognitive and sensory faculties. If Heidegger at certain points seems tempted by the Kantian task of identifying the existential conditions for the disclosure of beings as they are, he ultimately refuses to engage in the same a priori determinative investigation, turning to freedom or the open, or later, the clearing, as the necessarily incoherent origin of coherence, the nothing or silence that precedes concept and word—narrative itself.

3 Genuine Thinking

The *Contributions to Philosophy* (1936–1938) provide a clear formulation of Heidegger's turn away from transcendence as grounded in a specific being or structure to a relation to things that cannot be reduced to any set of principles:

> In philosophy, propositions are never subject to proof. This is so not only because there are no *highest* propositions, from which others could be derived, but because here "propositions" are not at all what is true, nor are propositions simply that about which they speak. All "proving" presupposes that those who understand, as they come to stand before the represented content of the proposition, remain the same, unaltered in following the representational nexus that bears the proof. And only the "result" of the course of the proof can require a changed mode of representation or, rather, require the representing of something previously unheeded.
>
> In philosophical knowledge, on the contrary, the very first step sets in motion a transformation of the one who understands, and this not in the moral-'existentiell' sense, but rather with respect to Da-sein. In other words, the relation to beyng and, ever prior to that, the relation to the truth of beyng are transformed in the mode of the transposition (*Verrückung*) into Da-sein itself. Because, in philosophical knowledge, in each case everything is transformed at once—the being of humans into its standing in the truth, the truth itself, and thereby the relation to beyng—and because, accordingly, an immediate representation of something objectively present is never possible, philosophical thinking will always seem strange (*befremdlich*).
>
> Especially in the other beginning, the leap into the "between" must be carried out instantly—in pursuit of the question of the truth of beyng. The "between" of Da-sein overcomes the χωρισμός ["separation"] not by slinging a bridge between beyng (beingness) and beings as if they were two objectively present riverbanks but by transforming together, into their simultaneity, both beyng and beings. The leap into the "between" is what first reaches and opens Da-sein and does not occupy a ready-made standpoint.[18]

[18] Martin Heidegger, *Contributions to Philosophy (Of the Event)*, trans. Richard Rojcewicz and Daniela Vallega-Neu (Bloomington: Indiana University Press, 2012), 13–14.

Heidegger announces what amounts to a revolutionary attitude to transcendence. Rather than fixing a relation between Dasein and things as between Beings and beings, as the being that relates to Being, and defining that relation in a final or determinate way, Heidegger creates the basis for a relation that cannot be resolved into any final determinate form. Moreover, and perhaps more importantly, he emphasizes the intense dynamism of the relation: Dasein is no bridge, no mediator between two basically determinate or fixed sorts of being, each opposed to the other in their ontological difference. To the contrary, Heidegger advances a "position" that cannot congeal into a position because it is always changing—"everything changes at once"—such that any one interpretation of the change is limited merely to the degree it fixes the relation or attempts to determine it finally. Let us linger on this point for a moment since it has particular relevance to Fedorov's project of fixing and recuperating identities, individual human beings, as discrete beings that are also somehow interconnected in fixed and determinable ways.

Transcendence, as a relation to Being and beings, has typically been anchored in a particular being, such as God or the human subject. Heidegger eliminates this anchor. But he does so most formidably by eliminating the underlying claim for primacy in determining the other components of the relation. There is no way to come to an objective or final account of the relation because each of the components of the relation, Being, beings and Dasein, is transformed in relating to the other. Dasein cannot "jump over its own shadow" to see itself or the others from "outside itself." The thinking Heidegger holds to be truly philosophical is one that in each act of thinking offers transformation because each act of thinking, at any given moment, both reshapes the thinking being and what is thought: each act of thinking, like a work of art, is familiar yet recognizably foreign and unique insofar as it cannot simply be reproduced as it is in continuous, reliable repetition.

This aspect of Heidegger's mature thinking about transcendence is extremely important. Thinking is no longer the establishment of norms that by definition must be regular and repeatable. For Heidegger this kind of thinking lies on the surface as "mere" correctness, and its main purpose is to establish norms that can be adopted by all in continuous and reliable repetition. This is the thinking that provides us with the notion of the machine, and, indeed, this is thinking as essentially mechanical. To such thinking, Heidegger opposes a thinking which resists reproduction and continually starts out anew, blazing inimitable trails that turn into other trails. This thinking offers up no definitive norms, no final account, no discrete beings. On the contrary, this thinking reveals the degree to which the mechanical process of thinking refuses to think by ensuring that all thought is a "calculus" or grammar of repeatable formulae based on a fixed understanding of thinking or, if you like, on an invariable set of governing axioms.

In this respect, Heidegger likens genuine thinking to poetry, not because, as is often understood, thinking creates its own mythology (though he might resort to myth-making from time to time) but because thinking is in every moment of its own activity—if it is genuine thinking—like a work of art creating something that stands unrepeatable, unique and individual. The *Contributions to Philosophy* is itself this kind of work. Heidegger asserts this on the very first page when he makes the succinct point that: "Future thinking is a *course* of thought, on which the hitherto altogether concealed realm of the essential occurrence of beyng is traversed and so is first cleared and attained in its most proper character as an event."[19] "Course of thought" translates *Gedanken-gang* which Heidegger hyphenates, we may presume, to underline its dynamic character, a character, that is, neither congealing into any given norm, nor setting norms, but rather inaugurating a kind of thinking that is in each case *sui generis*, an event, that ties or gathers Dasein, Being and beings together in a specific way that cannot be *the* way or is *the* way only as what cannot be *the* way. This rather precious formulation is nonetheless important because it is characteristic of Heidegger's thinking to deny its own authority as somehow correct or true in a final manner while using the traditional, absolute assertoric structure of the tradition. Hence, the bald contradiction that emerges in claiming that thinking cannot be final is brushed off by Heidegger as are restrictions imposed on thought by logic which insists on finality and unequivocal authority.[20]

[19] Ibid., 5.

[20] Thus Heidegger not infrequently dismisses the law of non-contradiction itself as inherently problematic, as formal, as mere surface adjudicating (questionably) between different claims for correctness while avoiding their ground. The basic problem with absolute statements is that they make a claim to be correct that Heidegger suggests cannot ever be fully justified without reference to the particular framework in which they function. Accordingly, an absolute statement in regard to a philosophical position for Heidegger cannot itself be philosophical other than in a surface manner because it conceives of philosophy as merely another framework establishing truths whereas philosophy questions all such frameworks and itself as well.

4 Historie and Geschichte

If we transpose the rigid distinction Heidegger draws between thought as repetition and thought as unrepeatable onto a narrative such as deification, of an approach to a final end, we come to the ground of the distinction between Historie and Geschichte. Like the distinction between truth as correctness and truth as disclosure, Historie describes a narrative structure that is normative to such an extent that it is no longer taken as having any origin at all—it is a narrative that has become "naturalized" so that it seems not to be a narrative but the very way of things itself. Geschichte unsettles such a naturalized narrative, and arguably more radically, it unsettles the concept of narrative as such insofar as this concept understands narrative primarily as a normative model. And this is no incidental normative model of course but one that shapes all our activities in the world, both in terms of grasping them as actions in time and as actions that have specific teleological significance. This point may be made effectively initially in terms of causation.

There is a kind of thinking that attributes a cause to any given action but, in so doing, attempts to avoid inscribing that action into a greater teleological whole.[21] This is an "apostate" application of the principle of sufficient reason that affirms "nothing is without a cause" (*nihil est sine ratione*), while not necessarily affirming the latter half of the phrase "why it is rather than not" (*cur potius sit quam non sit*). Separating these two statements allows one to insist on causal determination of everything that happens without presupposing a final end or pattern to events: an all-encompassing narrative or identity. Baruch de Spinoza's thought is one of the best examples of an insistence on causation that avoids the notion of an all-encompassing teleology. In the *Ethics*, Spinoza affirms the principle of sufficient reason as governing all events[22] but he denies that we can come to grasp the overall pattern of those events: after all, if the whole, as substance, is infinite, such a final purpose cannot be identified

[21] This kind of thinking is associated with mechanistic models of scientific knowledge that eschew teleological structures, a shift Ernst Cassirer identifies with the collapse of Aristotelian science at the dawn of modernity. See Cassirer's Introduction to his four-volume German edition of Leibniz's works: G. W. Leibniz, *Philosopische Werke in vier Bände*, eds. A. Buchenau and E. Cassirer (Hamburg: Meiner, [reprint], 1996), 1/xxxix–xl.

[22] Baruch de Spinoza, *Collected Works of Spinoza*, ed. and trans. Edwin Curley (Princeton: Princeton University Press, 1985), 410. We refer to Part I, Axiom 3.

because it would constitute a limit counter to infinity.[23] Spinoza derides attempts to establish a final causal purpose, even that of expressing the perfect "rationality" of the natural system itself. Spinoza's point is that "rationality" itself is complicated by anthropomorphic hopes such that what is rational becomes equated with what is useful for us.[24]

Heidegger in turn makes a surprisingly similar claim insofar as the rationality ascribed to the world in the notion that "nihil sine ratione est" is comforting in the vulgar sense that it avoids the more disturbing possibility that there is no rationality in the world, no underlying or divine order. Referring to our preceding comments, this more disturbing possibility suggests that one cannot be sure that what one knows will continue to hold of the world in endless and secure repetition. The very essence of the rational has to be predictability: a rational relation is one that holds always and not just once or not at all. If we use the word "coin" in relation to a coin on the table, that relation should always hold, both determining a correct statement and an incorrect one. If there were no stability of reference or, more radically, of syntax itself, then there would be no way to secure any statement: language and truth as correctness are only *useful* to the extent they reliably secure a given "territory" in the world, the precondition indeed to world itself as what takes the inchoate, unpredictable and indeterminable and transforms it into the clear, predictable and determinable.

Historie describes this latter process, whereas Geschichte is precisely what unsettles or sabotages it by putting it in question. While Historie purports to determine the governing structure of history itself and thus a master narrative that acts as the model for all other narratives, Geschichte intervenes in this governing narrative as both a challenge and revolutionary provocation. Put more simply, Historie seeks to end debate about history or narrative itself by asserting an authoritative model of history—and thus narrative—that excludes any notion of narrative that does not comply with it. These notions cannot be merely incorrect or wrong, for to be so affirms the authority of the narrative model, but rather they must seek to collapse the model itself: in this sense, they must somehow

[23] Spinoza famously rejects the imposition of ends on events as anthropomorphic, that is, as looking at events in the whole as having the same character as human events, i.e. as having a discrete beginning and end. See *Ethics* Part I, Appendix in Spinoza, *Collected Works*, 439.

[24] Ibid., 546. Part IV, Definition 1.

come "from outside" the model while being recognizable within it. This is one of the most delicate points of the encounter between Historie and Geschichte since the latter must be recognizable within the former but as that which pulls the former apart or undermines its authority by showing what it has ignored or suppressed in order to attain to that authority. This ignorance or suppression is the narrative equivalent of what Heidegger considers to be the primary problem of the idea (ἰδέα).[25] The idea preserves one view of a being as the view that becomes associated with the being in all reference to it: the tree we see in our mind's eye is a caricature of the tree, a mere coordination of shapes, that applies to all trees but is unlike any one in its exact detail. But that is not the main difficulty for Heidegger. The effect of the visual cliché is in fact to suppress other ways of viewing or relating to the tree, and these can be many of course since the tree can have as many looks as there are purposes given to it, whether they be of the simplest utility or those of poetic and artistic representation.

Heidegger first expresses this distinction in sections 72–77 of *Being and Time*. He revisits the distinction extensively in the later 1930s, mainly in his unpublished writings of the time and also in a lecture course from 1937–1938. Once again, the *Contributions to Philosophy* provides a decisive discussion which we quote at length:

> Historie spreads the illusion of complete mastery over all reality insofar as it sticks to everything superficial and displaces the surface itself which it takes as the only sufficient reality. Historie, as implying an unlimited knowledge of all things, in all respects, and with all the means of presentation, as implying dominion over everything factual, leads to the exclusion of Geschichte. The more decisive this exclusion becomes, the more unrecognizable it is to those who are excluded.
>
> Historie in its preliminary forms, in its development into science, and in the leveling down and intelligibility of this science to common calculation, is utterly a consequence of metaphysics, i.e., a consequence of the *Geschichte* of beyng, of beyng as Geschichte. Thereby, however, beyng and Geschichte remain completely concealed, indeed they even withhold themselves in this concealment.
>
> *Geschichte is beyng as appropriating event (Er-eignis) and must receive the determination of its essence on that basis, i.e., independently of any*

[25] See Heidegger, *Contributions*, 206, 208–213.

> *notion of becoming or development, independently of historiological considerations and explanations.* Therefore the essence of Geschichte can also not be grasped through an orientation toward the historiological "object," the object investigated, instead of proceeding from the historiological "subject," the subject conducting the investigation. What is the object of Historie then supposed to be? Is "objective Historie" merely an *un*attainable goal? It is not a possible goal at all. Then there is also no "subjective" Historie. By its very essence, Historie is grounded on the subject-object relation. Historie is objective because it is subjective, and insofar as it is the latter, it must also be the former. Therefore an "opposition" between "subjective" and "objective" Historie is entirely senseless. All Historie ends in an anthropological- psychological biographism.[26]

Aside from revealing the awkwardness of translation, in terms of distinguishing idiomatically between Historie and Geschichte in English,[27] this passage emphasizes and clarifies a few additional important points: (1) that Historie is connected with the inclusion of all things into a comprehensive whole, thereby "mastering" all reality (*Wirklichkeit*); (2) that the dominion of Historie entails the exclusion of Geschichte; (3) that Historie is nonetheless dependent on Geschichte, on this exclusion; and (4) that Historie (as mastery) is grounded in the discrete separation of subject and object and thus in the subject-object relation itself.

The link between Historie and mastery (or dominion) ties Historie to logic, to truth as correctness, as we have noted; but it also ties Historie to technology and, perhaps more ironically, to the elimination of time itself. The tie to technology is repeatedly mentioned in the *Black Notebooks*. Historie interprets everything in accord with what is now taken for real and rejects any other possible interpretation. Moreover, like technology, Historie is concerned to "eternalize" all things by subtracting them from any unpredictable change and thereby assuring that nothing new or unexpected can ever take place: what we know now is true for all times, and in this respect time as a measure of change has been maximally reduced in something like the eternal return of the same. As Heidegger notes in his lectures on Nietzsche, the eternal return of the same is the most comprehensive manner of imposing being on becoming, that is, of eliminating all

[26] Ibid., 388–389 (translated modified).

[27] The two English translations of the *Contributions* struggle with the distinction.

becoming to potentially endless repetition of the same.[28] For Heidegger this imposition of being on becoming is nihilism because it effectively eliminates any other way of approaching beings. The irony of this application of the fraught term "nihilism" to the complete securing of beings through Historie as a final regime of organization, control and mastery is that it favors exactly what has passed for nihilism in the vulgar sense of a loss of norms through the crumbling of their foundations.

On the contrary, Heidegger advocates little less than *revolution* through the overturning or overcoming of Historie as mastery, a false mastery in the sense that the very being who seeks to master reality merely turns itself into a being or thing by doing so, undergoing the most complete and radical process of reification. In this specific sense, Geschichte describes a revolutionary impulse, a profound restlessness that breaks through the mastery of Historie to explore its origins once again in a new beginning that recognizes its liminal status as creating a narrative within the space opened up by challenging accepted norms while not yet imposing new ones or, most radically, by dispensing with them as vestiges of a "metaphysical" past we must "twist out from" (*verwinden*) on the way to this new beginning. The moment in which a human being opens itself to Geschichte, to "happening," it becomes Dasein and starts along an entirely different path that cannot be replicated or reduced to what has become: Dasein finds itself precisely in this liminal space between what has come before and what has not yet come and seeks to realize itself not by closing itself off, by attempting to replace one Historie or regime of norms with another, but by holding itself open, to employ the metaphor again, to the sort of differential approach we mentioned at the beginning of the previous section in this chapter.

Here Heidegger is somewhat less direct about an important implication: that Geschichte is a kind of violence, an act of destruction that deploys the resources of Historie to undermine and overcome it, these consisting of related procedures of interrogation and investigation. And this violence is not merely of a kind that offends against specific norms, a violence that may be understood as criminal within the context of those norms. Rather, this violence aims at the ground of those norms itself, seeking to undermine not only that normativity, but, more

[28] Heidegger, *Nietzsche Volumes I and II: The Eternal Return of the Same*, trans. David Farrell Krell (New York: HarperCollins, 1991), 202. "To *stamp* Becoming with the character of Being—that is the *supreme will to power*."

broadly, normativity itself. This totalizing aspect of Heidegger's notion of Geschichte as perpetually subversive of norms, and thus normativity itself, is its genuinely revolutionary aspect aimed at maintaining an openness resistant to closure: conflict, not peace, is the "father of all things" as Heidegger is wont to recall for us referring to Heraclitus's fragment 53:

> The *polemos* named here is a strife that holds sway before everything divine and human, not war in the human sense. As Heraclitus thinks it, struggle first and foremost allows what essentially unfolds to step apart from each other in opposition, first allows position and status and rank to establish themselves in coming to presence. In such a stepping apart, clefts, intervals, distances and joints open themselves up. In confrontation, world comes to be. [Confrontation does not divide unity, much less destroy it. It builds unity; it is the gathering (*logos*). *Polemos* and *logos* are the same.]
>
> The struggle meant here is originary struggle, for it allows those that struggle to originate as such in the first place; it is not a mere assault on the present-at-hand. Struggle first projects and develops the un-heard of, the hitherto un-said and un-thought. This struggle is then sustained by the creators, by the poets, thinkers, and statesmen. Against the overwhelming sway, they throw the counterweight of their work and capture in this work the world that is thereby opened up. With these works, the sway, *phusis*, first comes to a stand in what comes to presence. Beings as such now first come into being. This becoming-a-world is authentic history.[29]

"Authentic history" translates the German, "eigentliche Geschichte." Geschichte is the operative term here and it denotes not only that "open" or "world" of "nothing that noths" that we have discussed in this chapter as the basic context in which Dasein is but, most importantly, the revealing of that context through the work of "poets, thinkers and statesman," of those who refuse to accept Historie, to obey what is at any given time accepted as the proper configuration of things. No matter what the specific approach, these poets, thinkers and statesmen do violence. A point Heidegger underscores later on in this same lecture series (from 1935):

> The uncanniest (the human being) is what it is, because from the ground up, it deals with and conserves the familiar only in order to break out of it and to let what overwhelms it break in. Being itself throws the human

[29] Heidegger, *Introduction to Metaphysics*, 67–68.

> being into the course of this tearing-away, which forces the human being beyond itself, as the one who moves out to Being, in order to set Being to work and thus to hold open beings as a whole. Therefore the violence-doer knows no kindness and conciliation (in the ordinary sense), no appeasement and mollification by success or prestige and by their confirmation. In all this, the violence-doer as creator sees only a seeming fulfillment, which is to be despised. In willing the unprecedented, the violence-doer casts aside all help. For such a one, disaster is the deepest and broadest yes to the overwhelming. In the shattering of the wrought work, in knowing that the work is unfit and a *sarma* (dungheap), the violence-doer leaves the overwhelming to its fittingness. But none of this takes the form of "lived experiences in the soul," in which the soul of the creator wallows, and it is absolutely not a petty feeling of inferiority; instead, it occurs solely in the manner of setting-into-work itself. The overwhelming, Being, confirms itself in works *as history* (Geschichte).[30]

Geschichte shatters (*zerbrechen*) the "wrought work" that is Historie: Gechichte as the happening of the open, the nothing, of Being itself brings us to the *now* as the ever uncertain, puzzling and estranging. For there is nothing more estranging and more nonsensical than the now of which Aristotle can say only that it is not. The now is the genuinely open that we tend to cover over or exclude from our thinking because the now in and of itself seems stubbornly to elude thought. We do not and cannot speak of ourselves in the now *in* the now: the I who speaks loses itself in the act of speaking since our understanding lags behind, imposing a constructed duration upon us by the mere fact that, despite our earnest intentions, our understanding takes place in time and requires time or temporal order as a syntax: no syntax, no thought. Yet, is this not itself the most estranging thought of all, that we are cut off from the originary "experience" in the now by our necessary reliance on a syntax that extends beyond or *transcends* the now? Here transcendence takes on yet another meaning insofar as transcendence speaks for alienation and estrangement from the very beings to which it supposedly opens us. The transcendence of ourselves, the capacity to "go out" to beings that is characteristic of transcendence and the ever elusive open in which transcendence happens (*geschehen*)—transcendence and happening in this sense consisting of one and the same—is also an alienation or estrangement. For only by transcending ourselves does syntax first become perceivable, and the impulse

[30] Ibid., 181–182.

to fix and determine this syntax once and for all may well seem irresistible as a way of covering over or minimizing that alienation or estrangement.

Heidegger seeks to revive the sense of estrangement, the wonder and perplexity (or horror and *Angst*), that comes from recognizing that what seems nearest to us, our "now," our very being as "in the now," is in fact farthest from us, most elusive and difficult. More than that, the manner of thinking or perceiving ourselves that is central to our having any sense of ourselves reveals its sterner aspect: that our more general capacity to know ourselves and beings assumes and cannot avoid a greater, more consuming ignorance—in coming to ourselves as such, we estrange ourselves from ourselves as well. This peculiar estrangement is a *Verrückung* that distances us from ourselves in attempting to come to ourselves. We cannot overcome such estrangement but exist in a continuous "between" that may be identified as the struggle to secure the now which ends up in estranging us from the now; the most dangerous result of this struggle is not the struggle itself but rather the possibility of a more complicated estrangement that is the estrangement from estrangement itself. This latter estrangement, to which Heidegger otherwise refers as the forgetting of the forgetting of Being, solidifies itself in the regime of correctness, Historie and technology.

5 Technology and Nihilism

The impact of this thinking for the narrative of deification and technological transformation is profound and pervasive. Deification cannot help but end up in a static Historie unless Geschichte intrudes upon it. Historie is indeed the word for narrative itself understood as any repeatable structure of action. And who can live or exist without a narrative? Geschichte is not to be understood merely in opposition to narrative or as a kind of "anti-narrative" or "negative narration." Geschichte describes what must finally elude description if description cannot free itself from repeatable representation: Geschichte is yet another Heideggerian word for what must evade capture in narrative, in logic, in categorization, in any of the means of classification that confirm beings as beings and exclude Being itself because it cannot be made sense of in terms of beings. Geschichte is precisely this unassimilable other that, once assimilated, loses itself. Geschichte is the now on which narrative depends but which cannot be grasped as it is in narrative. Geschichte is the nothing that noths, nonsense to Carnap, nonsense to those who refuse to consider Being as having any impact on beings other than to confuse and muddy our picture of the world, as Heidegger puts it, "to remove the comprehensible into its impossibility."[31] Geschichte is then the ghostly other that remains latent in narrative and may intrude on narrative to put it in question, rendering it incomprehensible, strange and unsettling precisely on that account alone.

Narrative faces dissolution in the same originary violence from which it emerged, as language from silence. Yet, the persistent irony is that narrative cannot itself describe its own dissolution without betraying it and reasserting its mastery over dissolution. Hence, the problem that haunts our own narrative in this book: how to reveal the fissures in a narrative without simply re-affirming narrative, how to put a narrative in question and interrogate its basic formative principles without turning in an unending circle whereby what is "refuted" or resisted returns in the refutation or resistance itself? Heidegger saw this problem clearly and exhorts us not merely to refute but overcome since every refutation merely re-asserts what it claims to refute. There is a dependence of the refutation on what it refutes, and this is merely another way of saying, in the case of any narrative, that the narrative cannot describe its own dissolution or its other without re-assimilating it. If one is to question and to challenge fundamentally, proceeding into the precincts of the nothing that noths, does one not end up in facile contradiction or "self-consumption"? Heidegger does not seem to think so insofar as he suggests that one can speak of the nothing or one can speak silently in contrast to Carnap who dismisses such speech as nonsense because it is not logical speech. Unlike Carnap, Heidegger believes that language can speak in many ways, not just in a logical way: one can say something understandable but not logical or clear in Carnap's scheme. In other words, logic does not grasp every aspect of human life, a point that is crucial to Heidegger's understanding of modernity as oriented towards the elimination of any "nonsense" or remainder that cannot be logically clarified or codified in an algorithm.

Heidegger deals with these issues and their almost apocalyptic consequences with unparalleled intensity in his lectures on Nietzsche and, in particular, in the important lecture, "European Nihilism" from 1940. In this lecture, Heidegger gives an account of modernity as nihilistic that is nested in an interpretation of Christianity and Descartes:

[31] Heidegger, *Contributions*, 136 (translation modified).

> Those who treat of beings as a whole in this manner are "theologians." Their "philosophy" is philosophy in name only, because a "Christian philosophy" is even more contradictory than a square circle. Square and circle are at least compatible in that they are both geometrical figures, while Christian faith and philosophy remain fundamentally different. Even if one wished to say that truth is taught in both, what is meant by truth is utterly divergent. Medieval theologians' having studied Plato and Aristotle in their own way, that is to say, by reinterpreting them, is the same as Karl Marx's using the metaphysics of Hegel for his own political *Weltanschauung.* Viewed correctly, however, the *doctrina Christiana* does not intend to mediate knowledge about beings, about what the being is; rather, its truth is throughout the truth of salvation. It is a question of securing the salvation of individual immortal souls. All knowledge is tied to the order of salvation and stands in service to securing and promoting salvation. All history becomes the history of salvation: creation, the fall, redemption, last judgment. This itself determines the manner in which (that is, the method by which) what is alone worth knowing is to be defined and mediated. *Schola* ("schooling") corresponds to *doctrina,* and the teachers of the doctrine of faith and salvation are therefore "scholastics."
>
> What is new about the modern period as opposed to the Christian medieval age consists in the fact that man, independently and by his own effort, contrives to become certain and sure of his human being in the midst of beings as a whole. The essential Christian thought of the certitude of salvation is adopted, but such "salvation" is not eternal, other-worldly bliss, and the way to it is not selflessness. The hale and the wholesome are sought exclusively in the free self-development of all the creative powers of man. Thus the question arises as to *how* we can attain and ground a certitude sought by man himself for his earthly life, concerning his own human being and the world. While in the medieval world it was precisely the path to salvation and the mode of transmitting truth *(doctrina)* that was firmly established, now the *quest* for new paths becomes decisive.
>
> The question of "method" — that is, the question about "*finding the way,*" the question about attaining and grounding a certainty secured by man himself — comes to the fore.[32]

If the Christian promise of salvation remains uncertain as a future possibility that may come after death for those who have lived selflessly and have faith in Christ, the modern promise of salvation seeks to realize

[32] Martin Heidegger, *Volume IV: Nihilism* in *Nietzsche Volume III: The Will to Power as Knowledge and as Metaphysics and Volume IV: Nihilism*, trans. David Farrell Krell (New York: Harper, 1987), 88–89.

immortality in the world, to make it certain and secure. Modernity seeks divinity on earth not through self-abnegation, but through self-assertion; not through submission to God, but through human action and power in an attempt to make everything logically transparent and clear. To achieve complete clarity and assert itself as the final, correct regime of truth modernity must eliminate negation by eliminating death. In doing so, modernity "takes over" the Christian striving for freedom from death.[33]

In the transition from the Christian to the modern era, Heidegger views Descartes as playing an important role because Descartes enshrines "subjective freedom" as "the absolute and founding authority of negativity."[34] Freedom is the capacity to refuse or reject—I am I because I can say no. By putting forward "a radical principle of negative freedom," Descartes expresses the human egoism at the center of metaphysics as an anthropomorphism based on an underlying equation of the striving for will to power as negation of any resistance to will, including thought itself.[35]

These two points can be tied together instructively with reference to Fedorov (and the θέωσις tradition as well): (1) Fedorov enshrines human egoism and specific identities, even if collective, in his common task such that Geschichte is simply ignored; and (2) he pursues a Christian notion of freedom that, as essentially negative from Augustine to Descartes, turns out to be negative in the most radical sense as self-annihilating: Christian freedom is a self-annihilating anthropomorphism that in attempting to turn human beings into divine beings transforms them into "bodies emptied of spirit" in Kojève's terms; namely, beings that are no longer conscious of themselves and can have no more resemblance to human beings than any animal, plant or stone.[36] In negating the realm of transcendence, in Heidegger's terms, to a frozen identity, a static being,

[33] Ibid., 89. Translation modified. The German is *übernehmen*.

[34] Alain Badiou, *L'Un: Descartes, Platon, Kant 1983–1984* (Paris: Fayard, 2016), 16.

[35] Ibid., 16.

[36] Alexandre Kojève, *Introduction à la lecture de Hegel* (Paris: Gallimard, 1968), 387–388. Kojève writes: "The end of History is the *death* of Man as such. After this death there remain (1) bodies living having human form but emptied (privés) of Spirit, that is, of time or creative power; and (2) a Spirit that exists empirically but in the form of an inorganic reality, one not living: as a book that, not having animal life, has nothing to do with time" (Our translation). It is unfortunate that this passage is not included in the widely available English translation.

Fedorov pulls human beings out of time and space, out of all contexts such that it is difficult to understand what remains of them other than remains or ruins of a once abundant, dynamic world.

This claim may seem exaggerated, but a careful reading of Heidegger's lecture clarifies the connection between salvation, will to power, and negation in a way that shows how the negative cannot be eradicated or ignored in the pursuit of ostensibly positive objectives but is in fact of the essence of the striving to attain them. The pursuit of universal salvation ends up in annihilation; far from bringing us to an eternal present, ostensibly positive objectives bring us to an absence or silence that are indistinguishable from death. Here is Heidegger on the essence of power:

> Every power is power only as long as it is more-power; that is to say, an increase in power. Power can maintain itself in itself, i.e., in its essence, only if it overtakes and overcomes the power level it has already attained, we say: *overpowering*. As soon as power stalls at a certain power level, it immediately becomes powerless. "Will to power" does not mean simply the "romantic" yearning and quest for power by those who have no power; on the contrary, "will to power" means the empowering of power by power for its own overpowering.[37]

Power aims necessarily at its own overpowering; each power level is only an epiphenomenon of a process of constant self-overpowering. If this process is not infinite, it ends up in annihilation of the power and the agent wielding that power. If it is infinite, then it is constant movement that offers no stoppage other than as weakness. In either case, the notion that power leads to a recuperation of all human beings, a stilling of time, in a fixed and eternal present, is incoherent because it is precisely this fixed "reality" that renders power powerless—it can only be a final point or an end to movement, history. If it is not an end, thus the cancelling of the process in its completion, then it cannot be anything else then a moment in the continued self-overcoming of power through itself. It is perhaps all too easy to miss the radicality of Heidegger's point: all stoppage, all retardation of the flow of things, of power, is a moment of weakness, an ebb in the flow of power that must either be an end or merely a brief moment or retardation or cessation of motion.

[37] Heidegger, *Volume IV: Nihilism*, 7.

In *Introduction to Metaphysics*, Heidegger seems to refer to this self-overpowering power as the "overwhelming sway" (*das überwältigende Walten*) against which human beings struggle, succeeding for a time in resisting the sway and then failing. Heidegger develops this rather mythic account of the struggle to arrest the flow of things within the context of a tragic play, Sophocles' *Antigone*, effectively developing a myth inside another myth with the aim of describing the tragic position of human beings who exert all their efforts to arrest a power that overwhelms them and sweeps them away into decline and death. No matter how the human being tries to overpower this power, the latter wins out in the end. For, if the human being were truly capable of arresting this power, the human being would also fail to preserve itself if its life is indeed truly lived in motion and not in the retardation of motion. If motion is perceived in relation to time, we may argue that the attempt to arrest motion is at once also an attempt to arrest time. But to arrest time is to eliminate the element in which human life takes place or transpires—is there any life outside of time?

By asking this question, we perhaps arrive at the heart of the matter: Heidegger views the pursuit of power through technology as destructive of time and death, and after 1945 he critiques technology as pernicious not so much out of an environmental concern for nature but rather out of respect for time and death. For Heidegger there is nothing permanent in us, no soul, no essence, no beyond. We are thus nothing more than a living death, nothing more than bubbles in a river: "On flows the river ceaselessly, nor does its water ever stay the same. The bubbles that float upon its pools now disappear, now form anew, but never endure long. And so it is with people in this world, and with their dwellings."[38] All flows, all is becoming; all that exists has the seed of death germinating in its very birth.[39] If the acorn reaches its purpose when it grows into an oak tree, we reach our purpose when we die. Yet, we often turn away from our purpose, from our essence as beings fated to die, by escaping into many and various fantasies of freedom, none of which, in our modern era, seduces us more than technology. For technology might allow us to

[38] Kenkō and Chōmei, *Essays in Idleness* and *Hōjōki*, trans. Meredith McKinney (New York: Penguin, 2013), 5.

[39] Our formulation partly comes from Hegel who says something similar in the beginning of the *Logic*. G. W. F. Hegel, *The Science of Logic*, trans. George Di Giovanni (Cambridge: Cambridge University Press, 2010), 60.

realize the Christian promise of freedom from time and death. If Fedorov would welcome such freedom, Heidegger would not because freedom from death, if it ever were to succeed, would terminate the human as a being distinctively capable of death. Now, it is not clear if Heidegger ever thought that technology could one day successfully overcome death. In *Introduction to Metaphysics*, he suggests that death is a limit that cannot be surpassed. "The human being has no way out in the face of death, not only when it is time to die, but constantly and essentially. Insofar as the human *is*, each stands in the no-exit of death."[40] But it is clear that he viewed technology as evasive of time and death, as having no respect for the ephemeral flow of being in the world.

A possible weakness in Heidegger's critique of technology, if it is a critique,[41] might be that it relies on a preference for restraint, or, in a more Chinese turn of phrase, for harmony. But can technology be harmonious? What is harmony? Is the pursuit of power unharmonic for nature? How do we know? And, if it is, what is wrong with disharmony? What is wrong with overcoming the human? Why should Dasein be preserved? Does Heidegger ever address these questions directly?

[40] Heidegger, *Introduction to Metaphysics*, 176.

[41] We hesitate because the tone of Heidegger's writings on technology varies from one text to the next. In his lectures on Nietzsche, Heidegger's tone seems prophetic; in his *Country Path Conversations*, it seems resigned; in his Bremen and Freiburg lectures, it seems critical.

CHAPTER 4

Reconciliation—The Great Harmony

Heidegger again:

> If communism in *China* should come to rule, one can assume that only in this way will China become 'free' for technology.[1]

If China becomes free for technology, if it fully embraces Western modernity, then there will be no cultural power left to question and moderate, if not resist, the global ascendancy of Western modernity. There will be no hope left for humanity, no country to save it—not Germany in which Heidegger saw hope in 1933; not Russia in which he saw hope in 1939–1941; not China in which he saw hope in 1945.[2] Humanity will be "done for" without Germany, Russia, and China because there will be no philosophical or "*questioning*" culture left to stop the unceasing striving for power that constitutes Western technological modernity, which, so

[1] Martin Heidegger, *Anmerkungen I–V (Schwarze Hefte 1942–1948)*, GA 97 (Frankfurt am Main: Vittorio Klostermann, 2015), 441.

[2] Heidegger showed a sustained interest in Russian culture going back to 1908. See Martin Heidegger, *Überlegungen XII–XV (Schwarze Hefte)*, GA 96 (Frankfurt am Main: Vittorio Klostermann, 2014), 131, 148, and 257. As to China, Heidegger mentions Zhuangzi in his *Country Path Conversations*, trans. Bret W. Davis (Bloomington: Indiana University Press, 2010), 157. Moreover, Heidegger attracted considerable interest in China and Japan. See Yuk Hui, *The Question Concerning Technology in China: An Essay in Cosmotechnics* (Falmouth: Urbanomic, 2016).

J. Love and M. Meng, *Revolutionary Bio-politics from Fedorov to Mao*,
https://doi.org/10.1007/978-981-99-4745-4_4

Heidegger thinks, will lead to our own overcoming, our own destruction: the technological tools that we created will one day overpower and destroy us.[3] If such a claim might sound histrionic or outlandish, none of us can deny that we appear to be close to the point of creating something very powerful and new with rapid advancements in areas such as artificial intelligence; in fact, AI is advancing so rapidly that even leading experts in the field have called for a pause to think about where we might be headed with it.[4] That the race to advance artificial intelligence in ever more powerful ways shows no signs of abating makes Heidegger sound more percipient than ever: "Every power is a power only as long as it is more power; that is to say, an increase in power. Power can maintain itself in itself, that is, in its essence, only if it overtakes and overcomes the power level it has already attained."[5]

Heidegger seems convinced that we have unleashed an unceasing, dynamic drive for power that threatens to overcome us. But is there really no hope of arresting the course of what we have created? Perhaps there is. Enter China once again, albeit this time through a thinker who, though his reputation has been blemished by the myopic politics of his conservative disciples in the United States, understands as astutely as Heidegger the revolutionary possibilities of technology:

> Man's humanity is threatened with extinction by technology. Technology is the fruit of rationalism, and rationalism is the fruit of Greek philosophy. Greek philosophy is the condition of the possibility of technology and therefore at the same time of the impasse created by technology. There is no hope beyond technological mass society if there are no essential limitations to Greek philosophy, which is the root of technology, to say nothing of modern philosophy. Greek philosophy was the attempt to understand the whole. It presupposed therefore that the whole is intelligible, or that the grounds of the whole are essentially intelligible and at the disposal of man as man — that they *are* always, and therefore in principle are always accessible to man. This view is the condition of the possibility of human

[3] Heidegger associates philosophy with the "power" to question everything, a power foreign to or incompatible with science or any other kind of knowledge that serves everyday needs. Heidegger, *Introduction to Metaphysics*, trans. Gregory Fried and Richard Polt (New Haven, CT: Yale University Press, 2014), 8–10.

[4] https://futureoflife.org/open-letter/pause-giant-ai-experiments/.

[5] Heidegger, *Volume IV: Nihilism* in *Nietzsche Volume III: The Will to Power as Knowledge and as Metaphysics and Volume IV: Nihilism*, 7.

> mastery of the whole. But that mastery leads, if its ultimate consequences are drawn, to the ultimate degradation of man. Only by becoming aware of what is beyond human mastery can we have hope. Transcending the limits of rationalism requires the discovery of the limits of rationalism. Rationalism is based on a specific understanding of what being means, viz., that *to be* means primarily to be present, to be ready at hand, therefore that *to be* in the highest sense means to be always present, to be always. … A more adequate understanding of being is intimated by the assertion that *to be* means to be elusive or to be a mystery. This is the Eastern understanding of Being. Hence there is no will to master in the East. We can hope beyond technological world society, we can hope for a genuine world society, only if we become capable of learning from the East, especially from China. But China succumbs to Western rationalism.
>
> Heidegger is the only man who has an inkling of the dimensions of the problem of a world society.
>
> There is needed a meeting of the West and of the East. The West has to make its own contribution to the overcoming of technology. The West has first to recover within itself that which would make possible a meeting of West and East: its own deepest roots, which antedate its rationalism, which in a way antedate the separation of West and East.[6]

Though Leo Strauss says "degradation of man," he also says, agreeing with Heidegger, that "man's humanity is threatened with extinction by technology" because technology seeks to overcome all limits and conditions. It seeks to make what is conditioned unconditional, what is changing unchanging, what is embodied disembodied. If technology aims to negate the human being as conditioned, changing, and embodied, if technology as will to power aims to negate all obstacles to its own expression, why would we want to advance it? Why would we want to create a disembodied intelligence that could outstrip and potentially destroy us? Where is the logic in destruction? Why the atomic bomb? Why destroy the very ecosystems on which our lives depend? Why destroy the different cultures and histories of the world in a modernizing rush to make everything the same? To repeat: Where is the logic in destruction?

If we follow what Strauss suggests in the above passage, there is a logic to it, and it lies in the logic of Western thought itself, specifically

[6] Leo Strauss, "An Introduction to Heideggerian Existentialism," in *The Rebirth of Classical Political Rationalism: An Introduction to the Thought of Leo Strauss*, intro. Thomas L. Pangle (Chicago: The University of Chicago Press, 1989), 42–43.

in the Western orientation towards permanence and stability. In other writings, Strauss offers slightly different answers, but he does not deviate from his focus on identifying intellectual developments in Western history as creating the condition of the possibility of modern technology. Hence, the importance of the Chinese tradition as an alternative to Western development for Strauss and, more clearly, for modern Chinese philosophers. Yet, presenting China as an alternative to the West, as Heidegger suggests, runs into the obvious problem that China has embraced Western modernity in a bid to catch up with, if not surpass, the United States as a global economic and technological power. In light of China's modernization, it is perhaps questionable whether its "Eastern understanding of Being" still exists, and, even if it does, whether it can be accepted or even heard in a West dominated by a rational, scientific understanding of Being, let alone whether ancient Chinese thought can be translated into a Western, twenty-first century language such as contemporary English.[7]

The obstacles are no doubt significant, but we do not wish to be pessimistic. Rather, we wish to explore the possibility of "reconciling" the East with the West, and, in doing so, the possibility of finding a harmonious technological modernity that rejects both deification and its opposite, Heideggerian finitude. We proceed as follows. After briefly discussing Mao's conception of revolutionary practice and contradiction, we turn to Mou Zongsan and contemporary philosopher Yuk Hui as offering two comprehensive responses to Mao and Western modernity more broadly through their creative interpretations of ancient Chinese thought. While Chinese thought in the twentieth century is extraordinarily rich and varied, we consider both Mou and Yuk as exemplary figures in terms both of their attempt to integrate western thought into the Chinese context and their concern with ascertaining a distinctly new attitude to science and technology.[8] Finally, we conclude by asking two basic

[7] Of these two obstacles, we might hazard to suggest that the latter obstacle may be slightly less formidable than the former since the tremendous historical variety of the English language supports creative and imaginative translation, as Michael Cooperson's translation of Al-Ḥarīrī's *Impostures* has recently shown. Cooperson translates Al-Ḥarīrī's Arabic in fifty different varieties of English to "reflect" his interpretations of Al-Ḥarīrī's literary and linguistic innovations.

[8] The variety of modern Chinese response to technology is of course enormous. But we may identify a basic, and arguably obvious, divide between the insistence on catching up to the west that distinguishes communist attitudes to technology and more conservative attitudes that had much to do with preserving the traditions of Chinese culture which

questions: (1) Can there be a meeting of East and West?; and (2) Can technology be harmonious with nature?

1 Mao Zedong

No single figure exercised a greater influence on Chinese history in the twentieth-century than Mao Zedong (毛泽东, 1893–1976). Mao's influence was of course not confined to China, and various "Maoisms" continue to play a significant role in contemporary politics and philosophy: the noted French thinker Alain Badiou remains an avowed Maoist and his comments on Mao and the infinite are of interest in grasping Mao's philosophical legacy aside from his legacy in politics and international affairs. In this section of our chapter, we focus on Mao's attitude to

had collapsed in the first decades of the twentieth century. Thus, one finds a ready divide between Maoists and Confucians or "new" Confucians (among whom Mou is a leading thinker). But there are other important figures who do not fit this basic divide so easily like Li Zehou who is Marxist but abhors the "voluntarism" he identifies with Mao. While Confucianism has become increasingly important in recent Chinese thought—Xi Jinping is fond of quoting the important Confucian thinker Wang Yangming (王阳明, 1472–1529)—other currents persist, like the "new" left thinking of Wang Hui. Our selection highlights two thinkers who seek to create a renewed Chinese tradition capable of global influence that speaks to the preeminently philosophical concerns about technology articulated by Heidegger and many others. There are other notable "new" Confucians who held similar views like Thomé H. Fang (Fang Dongmei, 方東美, 1899–1977) and Tang Junyi (唐君毅, 1908–1978), but Mou is arguably the dominant figure in that "movement" and a highly creative and original philosopher in his own right; he also is an important reference point for Yuk. The "new" Confucian thinking and its offshoots oriented to harmony are also quite congenial (however ironic that may be) to the current policies of the communist party and Xi Jinping—the attempt is underway to create both a new attitude to technology that works harmoniously with the environment and an economic system that frees the world from the hold of a violent, even nihilistic, capitalism. Wang Hui (汪晖) has emphasized the importance of cultivating a Chinese approach that can be universalized while attaining to the revolutionary spirit diminished in the 1980s and 1990s. See Wang Hui, *China from Empire to Nation State* (Cambridge, MA: Harvard University Press, 2014) and two recent articles: "Twentieth-Century China as an Object of Thought: An Introduction, Part 1 The Birth of the Century: The Chinese Revolution and the Logic of Politics," *Modern China* 46, no. 1 (2020): 3–48 and "Twentieth-Century China as an Object of Thought: An Introduction, Part 2 The Birth of the Century: China and the Conditions of Spatial Revolution," *Modern China* 46, no. 2 (2020): 115–160. Wang Hui's project seems far more politically oriented than philosophical, his conceptual armature far less rich and well-articulated than those of Mou and Yuk despite the sheer volume of his writings.

the realization of the utopian state and the attitude to technology appropriate for such a state. Mao's thinking regarding the purpose and practice of revolution differs markedly from the Soviet precedent. Although Mao's thinking shows affinities with Fedorov, Mao does not envisage an end of history as a final point in time when time is recuperated but rather as the establishment of a revolutionary notion of infinite striving that has more affinity with cyclical thinking than the sort of rigid linearity that distinguishes Fedorov and the tradition from which he springs.[9]

Mao lays out the prime contentions of his revolutionary thinking in two fairly brief and celebrated essays from 1937, entitled "On Practice" (实践论) and "On Contradiction" (矛盾论).[10] "On Practice" gives an account of the theory of knowledge relevant to the revolutionary order, whereas "On Contradiction" gives an overview of what we may refer to, somewhat extravagantly perhaps, as the more sweeping metaphysical view that grounds Mao's revolutionary practice. In this respect, the two essays complement each other and provide us with a broad account of the thinking that in 1949 would rise to dominance in China. While Heidegger's concern with China's becoming merely another technological colony of the West is not unwarranted, an examination of Mao's essays should complicate the picture considerably as well as offer a bridge to the divergent and yet complementary concerns of his successors and opponents. In a word, we can suggest that Mao regards technology as part of a process that is not necessarily opposed to nature but rather is an expression of it. Mao's thinking, so important in China, does not

[9] We suggest, as have others, that Mao is a thinker of perpetual revolution. The moment the revolution reaches a point of stasis, it is over and has lost its way. Mao has been called by both Slavoj Žižek and Jonathan Spence a "Lord of Misrule" on this account. For our purposes, there is an intriguing homology between Mao's notion of perpetual revolution and Heidegger's description of power in his Nietzsche lectures where Heidegger affirms the notion that power declines the moment it does not overcome itself.

[10] Mao Zedong, *On Practice and Contradiction*, intro. Slavoj Žižek (London: Verso, 2017), 52–102. Mao's thinking has also not received as much attention as it deserves in English. There are a few studies worth consulting: S. Schram, *The Thought of Mao Tse-tung* (Cambridge: Cambridge University Press, 1989); Robert Elliott Allinson, *The Philosophical Influences of Mao Zedong* (London: Bloomsbury, 2020). Mao is of course not considered a philosopher, but his thinking is far from philosophically uninteresting or "amateurish" (a rather impoverished insult employed on occasion), and it had in various ways an enormous influence in China and elsewhere, not only in the truncated form of "Mao Zedong Thought" (毛泽东思想) but also in the form of the short talks from 1937 and, especially, in his treatment of guerrilla warfare.

necessarily fall prey to Heidegger's critique—or at least it can be so interpreted—and this opens up for us a fundamentally different picture of technological change than Heidegger's, one indeed whose closer affinity in Western thought lies not with Marx but with Spinoza (and one that therefore avoids rebelling against modernity through either Romanticism or Fascism).

Mao writes: "Marxists recognize that in the absolute and general process of development of the universe, the development of each particular process is relative, and that hence, in the endless flow of absolute truth, man's knowledge of a particular process at any given stage of development is only relative truth. The sum total of innumerable relative truths constitutes absolute truth."[11] Here Mao expresses a view that emphasizes an underlying unity of human activity with the cosmos (universe), hardly a view that indicates enmity or that could be enlisted in support of a mastery or regulation narrative in the sense of Fedorov. There is no "blind force" to regulate and correct. To the contrary, Mao asserts a kind of harmony, but of a distinctive sort, as the essay "On Contradiction" indicates. Before we delve into that essay, we should bring out a few more salient aspects of "On Practice."

As the title indicates, Mao foregrounds the practical orientation of knowing and makes the important claim that knowers who lack practice are not genuine knowers. "Many theories are erroneous and it is through the test of practice that their errors are corrected. That is why practice is the criterion of truth..."[12] Practice as "criterion of truth" is crucial, and the relation of theory and practice becomes reversed in Mao: theory becomes a kind of practice. It becomes reversed in Mao because Mao asserts through the primacy of practice the same equality of the learned and the unlearned that is such an important aspect of Fedorov's thinking. This equality is all the more significant because it acts as a fundamental orientation of the Cultural Revolution, however terrible the results of its attempt to close the gap between the learned and unlearned became. The central principle is clear: the unified effort to achieve equality and well-being for all through knowledge and technology is the imperative of practice and a result of practice. Moreover, as we have noted, this practice does not constitute an abuse of nature but rather an *expression* of it

[11] Mao, "On Practice," 64.

[12] Ibid., 62.

or of the universe of which we form an active part. We develop and, in so doing, develop the universe. We do not bring nature to a stop but reveal our own nature in developing ourselves *within* nature.

As "On Contradiction" tells us, these developments are dialectical: development and growth take place in relation to other things that are also growing and developing: "As opposed to the metaphysical world outlook, the world outlook of materialist dialectics holds that in order to understand the development of a thing we should study it internally and in its relations with other things; in other words, the development of things should be seen as their internal and necessary self-movement, while each thing in its movement is interrelated to and interacts on things around it."[13] Mao continues: "[t]he fundamental cause of the development of a thing is not external but internal; it lies in the contradictions within the thing."[14] All things are related to each other as developing things and their development is driven by contradictions. Hence, the basic motive force in things are contradictions. If things are variable in the sense that the wholes in which they develop are multiple and variable, contradiction proves to be a transitive principle of development that functions consistently throughout the whole.

Unlike properly Hegelian or Marxist development, there is no final end to this development. The rejection of an end is surely one of Mao's most far-reaching innovations: the promise of utopia remains as a promise and is indeed only realizable in an infinite striving for full self-realization that never comes to its end if that end means the final banishment of contradiction. To banish contradiction would be to banish development and arrive at a static point that is the refutation of all growth (if such growth is driven by contradiction) and an end that must be inseparable from death or dogmatism, the terrible rule of an irrefutable authority.

Let us now combine the two essays with their concluding points. "On Practice" concludes strikingly: "The form repeats itself in endless cycles, and with each cycle the content of practice and knowledge rises to a higher level. Such is the whole of the dialectical-materialist theory of knowledge, and such is the dialectical-materialist theory of the unity of knowing and doing."[15] Mao asserts a cyclical development of knowledge

[13] Mao, "On Contradiction," 69.

[14] Ibid., 69.

[15] Mao, "On Practice," 66.

that would be familiar within a Hegelian context. But he adds in "On Contradiction" that "the struggle of opposites is ceaseless, it goes on both when the opposites are coexisting and when they are transforming themselves into each other, and becomes especially conspicuous when they are transforming into one another; this again is the universality and absoluteness of contradiction."[16] These assertions, taken together, create a potent cosmological vision of the universe as a dynamic, essentially unfinished, system ruled by struggle and contradiction whereby differences are resolved and re-affirmed such that there is never a final moment of reconciliation. Struggle is, moreover, not inherently violent but rather the interplay of beings that announces a deeper transformative structure whereby beings merge into and become each other. Mao is careful to affirm the unity created by opposition:

> We Chinese often say, 'Things that oppose each other also complement each other.' That is, things opposed to each other have identity [are identical to one another]. This saying is dialectical and contrary to metaphysics. 'Oppose each other' refers to the mutual exclusion or the struggle of two contradictory aspects. 'Complement each other' means that in given conditions the two contradictory aspects unite and achieve identity. Yet struggle is inherent in identity and without struggle there can be no identity.
>
> In identity there is struggle, in particularity there is universality, and in individuality there is generality.[17]

This struggle without resolution is not the struggle of perpetually alienated enemies. The struggle for identity is in fact the struggle to express the generality and communality that is initially hidden in the relationship—indeed, that is the reason for struggle itself. If we take this passage in combination with the others immediately above, we may come to a bold assertion: that struggle emerges from a *misrecognition* of underlying communality, and the result of struggle is not continued enmity but rather the dynamic unfolding of the generality and the universality—of the whole itself—that we come first to understand, from our given point of view, as separate.

Now we may return to "On Practice" and its apparently simple relation of perceptions and conceptual wholes. Mao notes that perceptions identify

[16] Mao, "On Contradiction," 101.

[17] Ibid., 99.

separated things and bring them together according to certain points of view. Knowledge results when those points of view are combined and begin to bring to the fore the various interrelations of the perceptions (which at the outset seemed separate). Knowledge continues to grow as these interrelations expand and develop. We may understand that they expand and develop through the clash of different views in so far as these different views insist on their independence and reject interdependence. The dialectical view accentuates interdependence and weaves together the various views in what we may think of as an ever-ascending whole. But this whole ascends as it widens—both the vertical and horizontal aspects of knowing turn out to be unified as well so that each reflects the other and is an imitation of the other from a specific orientation. The crucial point to grasp is that the development of knowledge and practice is an expression of the self-differentiating whole that comes to increasing self-realization as contradictions are resolved and emerge anew in the cyclical becoming of the whole that is *endless*. As Badiou had affirmed, Mao thinks infinitely because he thinks of a whole without end, of a whole that is never whole.

There have been complaints about the weakness of Mao's Marxist understanding or his unusual Hegelianism or his bringing together heterogeneous elements.[18] Yet, as we shall see, Mao is thinking in an eminently Chinese way, and his adaptation of Marxist ideas is a not unsophisticated Sinicization of those ideas that in effect transforms them in accord with distinctively Chinese habits of thought. Most important of all for our immediate purposes is Mao's twofold denial: (1) of the irremediable hostility between technical progress and nature; and (2) of the possibility of an end to history. We argue in the coming sections that, far

[18] See, for example, Wang Fanxi's critical comments on Mao: Wang Fanxi, *Mao Zedong Thought* (Leiden: Brill, 2020), 172–198. Wang Fanxi (王凡西, 1907–2002) was a leading Chinese Trotskyist and his views are colored by his dismay at Mao's ascendence but he is more even than many Western observers, less inclined either to dismissal or inordinate praise. Wang notes in chapter 7 of his book (pp. 172–193) the weakness of Mao's Marxist education and the influence of traditional Chinese—especially Confucian—elements in Mao's thought despite Mao's later rejection of Confucianism. Wang does not dismiss Mao's thinking, however; he simply emphasizes its essentially practical character, not as an insult, but as an observation that, for Mao, theory served practice. As we noted previously, others, like Žižek and Spence, emphasize an intriguing account of Mao as a "Lord of Misrule" constantly destabilizing and changing. As such, we speculate about the possible subterranean influence of Zhuangzi, surely an unusual conclusion and one wholly counter to the earnestness typically associated with revolutionary ardor.

from being counter-revolutionary positions, these carefully-crafted Maoist positions offer a revolutionary retort to Fedorov *and* Heidegger.

2 Mou Zongsan

Mou Zongsan (牟宗三, 1908–1995) was violently opposed to Mao and the communist party; upon Mao's victory in 1949, Mou left mainland China for Taiwan and spent the balance of his life in Hong Kong and Taiwan. Among other things, Mou objected to the Western orientation of the communist party and to Mao's apparent disdain for the Chinese past and Confucian thought in particular. Yet, Mou too was a product of the same May-fourth movement that had ignited an entire generation, including the communists,[19] and his attitude to science and technological progress was not one of outright rejection, but rather nuanced and flexible. Like many other Chinese intellectuals of his generation, Mou sought to create a synthesis between Chinese and Western cultures and a new synthesis where Chinese culture could benefit the West while liberating itself from the perceived ills of western influence. Not unlike Strauss's Heidegger, Mou strove to create the conditions for a *rapprochement* between China and the West and he did so through a sprawling and absorbing philosophical project that is surely one of the most impressive of the twentieth century if still relatively little-known outside of the Chinese world.[20]

[19] We are referring to a crucial movement in modern Chinese self-determination initiated by student protests in Beijing that took place on May 4, 1919 as a protest to concessions given to Japan in the Treaty of Versailles. The consequences of the protest were broad and encouraged an upswing in Chinese nationalism affecting both communists and nationalists. It is an intriguing fact that both Mao and Mou (who, as we noted, was vehemently anti-communist and rejected Mao in the most aggressive terms) reflect the influence of this movement as nationalists and advocates of a China liberated from baleful Western influence and domination.

[20] We might caution, however, that Mou, like Heidegger, was a nationalist and saw his thought as a recovery of a distinctively Chinese tradition and attitude. There are still regrettably few studies of Mou's thought in English, but those we do have offer insights into Mou's concern with morality and, especially, his abiding interest in Buddhism and the philosophy of Immanuel Kant. See Jason Clower's study, cited below, as well as two other important studies: Sébastien Billioud, *Thinking Through Confucian Modernity: A Study of Mou Zongsan's Moral Metaphysics* (Leiden: Brill, 2012), and N. Serina Chan, *The Thought of Mou Zongsan* (Leiden: Brill, 2011).

In a word, Mou considered himself a Confucian and a renovator of the Confucian tradition.[21] Yet, he displayed a remarkable interest in most of the main traditions of Chinese thought, including Daoism and especially Buddhism. While some scholars consider this latter interest unusual, especially in a Confucian scholar, Mou's "Confucian" colleague Tang Junyi (唐君毅, 1909–1978) showed a similar orientation. In all his inquiries, Mou's thinking shows a singular concern to define sagehood and to do so in the context of the modern world and the terrible depredations that had done so much to dissolve the traditional structure of Chinese culture and thought. We might even be tempted to say that Mou sought to establish a notion of sagehood with global (or universal) application that fused the various Chinese traditions and connected them with Western thought through Kant who seems to have exerted an unusual fascination on Mou.[22]

For our purposes, Mou's thought is of interest since his notion of the sage seems to contrast strongly both with the tradition of θέωσις we discussed in the second chapter and that of the finite, ever alienated philosopher in the third. If Mao sought to bring forth a revolution not only in politics, but in thought that would open China to technological progress on an unprecedented scale, Mou's thinking offers a nuanced attitude reflecting a concept of wisdom, not as conquest of nature or endless questioning, but rather of reconciliation and harmony. While Mou's thought is far too complicated and ambiguous to classify easily, we see in it broad affirmation of a harmonious relation to the world and all other beings that seems foreign both to Fedorov and Heidegger, if not also to Mao, for whom struggle is still a central feature of human existence. Mou introduces us to quite a different reality in which human beings are not fundamentally alienated in a hostile world but constitute a living part of that world. In this respect, there is a similarity between Mou's fundamental attitude and that expressed by Mao which shows in both the orientation of Chinese culture to harmony and reconciliation with the world—conquest of nature or a despairing affirmation of our essentially alienated status in the world as creatures bound by a death they

[21] Mou is considered a "new Confucian" as opposed to the "Neo-Confucians" of the Song-Ming period (960–1644).

[22] Heidegger's presence in Mou's work should not be neglected either. If Kant is the major Western influence, Heidegger is surely a highly important one.

recognize are not defining features of Mou's thought or of the Chinese wisdom he attempts to revive and renew.

Given the vast scope of Mou's work, we have chosen to focus on his interpretation of Buddhism and the Buddhist sage as outlined in his *Nineteen Lectures on Chinese Philosophy* (中国哲学十九讲) delivered at the National Taiwan University in 1978 and available in English.[23] This approach has several benefits. While Mou considered himself a Confucian, some of his most provocative and exploratory theoretical work emerges in the context of his studies of Buddhism, and this is certainly the case with the *Lectures* where the core lectures, or what we consider the most absorbing ones, deal largely with Buddhist themes.[24] Moreover, Mou addresses technology directly in these lectures and links technology with the Buddhist notion of "expediency" (方便), a striking linkage that accepts technology while relegating it to an inferior status in terms of wisdom—that is, technology, far from being a means to deification is a form of attachment that effectively turns one away from genuine wisdom. For Mou technology is a way of getting on with the world that retards sagehood or indeed diverts one away from sagehood to the extent one becomes a technological being. While this is not an attitude of overt hostility or apocalyptic doom in the Heideggerian sense, it is also not an attitude that views technology as salvific.

Accordingly, we may view Mou's attitude to technology as profoundly ambiguous, and indeed it is, but for reasons that have to do with the form of Buddhism Mou prefers to examine, Tiantai Buddhism. Mou claims that

[23] There are two translations of these lectures available in English, one online, and one in book form. The online version is more scholarly but it can also be unclear and unidiomatic. The book edition has some similar faults but offers useful glosses on Mou's terse and compressed Chinese. Since we cannot readily recommend one translation over the other, we have taken the somewhat unusual decision to present the book edition first, followed by the online edition for longer set-off citations of the text. For shorter citations, we have used the print edition as a base. All citations have been checked against the Chinese original. The online edition is available at: https://www.nineteenlects.org. The printed edition is: Mou Zongsan, *Nineteen Lectures on Chinese Philosophy*, trans. Esther C. Su (San Jose, CA: Foundation for the Study of Chinese Philosophy and Culture, 2015).

[24] Jason Clower puts it well: "Mou believes that Confucians teach a style of practice which can produce top-level performers of a higher rank than Buddhists can, but that Buddhists traditionally devoted much more energy to theorizing the activity and made a number of major theoretical breakthroughs that made much better sense of it than the Confucians had done." Jason Clower, *The Unlikely Buddhologist: Tiantai Buddhism in Mou Zongsan's New Confucianism* (Leiden/Boston: Brill, 2010), 58.

Tiantai is the most perfect form of Buddhism,[25] and he creates a philosophical hierarchy of Buddhist teachings that combines a sophisticated two-truth attitude with a claim for universal Buddhahood in fact typical of the Mahāyāna tradition. Mou presents and expresses the basic orientation of his thinking compactly with the extraordinary phrase "one mind opening two gates" (一心开二门).[26]

Before proceeding to discuss the two-truth theory in detail, Mou sets out a basic distinction between West and East that recalls Strauss:

> Ordinary, worldly learning—exemplified by Western philosophy and tradition—"strives for being" (*struggle for being*). Buddhism, advocating just the opposite, "strives for non-being" *(struggle for non-being)*. Western science, philosophy, and religion focus on affirming "being." The emphasis on *being* in Western philosophy, which started with Plato and continued uninterrupted to Heidegger, illustrates the difficulties encountered in, and the great wisdom required for, grasping the concept of *being*. Since being is self-nature (*zixing* [自性]) and is "not emptied of substance," Buddhism strives to remove "being" by advocating that "[phenomena are] emptied of substance" (*xingkong* [性空]), "all dharmas have no self-nature" (zhufa wuzixing [诸法无自性]), and "all acts have no permanency, all dharmas have no substantial-self."[27]
>
> [As regards this difference, I have in the past made two pronouncements. In general the philosophies of the world—taking Western philosophies and the Western tradition as representative—have consisted of "the struggle for Being." This is characteristic of all Western sciences, philosophies, and religions. Being is simply self-nature [*zixing* 自性 , substance], in other words, substance [self-nature] is not empty. Ever since Plato, the emphasis has been on Being, and even in recent times Heidegger continues to explain Being, all of which shows that Being is very difficult to grasp. It is great wisdom to get a grip on Being. Buddhism is just the opposite of this, having been engaged precisely with "the struggle for Non-Being,"

[25] Mou, 327.

[26] The phrase can itself be traced back to the Huayan patriarch Zixuan (子璿) (965–1038) and figures prominently in the *Treatise on Awakening the Mahāyāna Faith* attributed to Aśvaghosa. See Clower, *Unlikely Buddhologist*, 119n89. Mou follows the Tiantai organization of Buddhist teachings into five levels beginning with the Flower Garland Sūtra and ending with the "perfect doctrine"—that of Tiantai itself.

[27] Mou, 262. We note the phrases and words italicized are in English in the original text.

with eliminating Being. If we eliminate being, then we are left with "emptiness of substance," which is to say, "All beings [dharmas] are without substance [諸法無自性]," "All karmas are without permanence, all dharmas are without self [諸行無常, 諸法無我]." Being without permanence and without self is simply to be without substance [self-nature], and to be without substance is emptiness.]

The basic opposition here is striking: Western thought "strives or struggles for being" (为实有而奋斗). Buddhism "strives or struggles to remove being" (为去掉实有而奋斗). Mou refers to these two struggles, in English in the original, respectively, as the struggle for being and the struggle for non-being. Mou thereby emphasizes radically different orientations between Western thought and Buddhist thought that come to full expression in the two-truth attitude. According to this attitude beings are empty of self-nature (自性) and empty of substance (性空)[28]; they are empty because they "originated from conditions." The two-truth attitude considers beings empty because they are *conditioned* and thus *impermanent*. But this attitude is unusual because it affirms both their being as beings in the "mundane" and "ordinary" sense and their emptiness as conditioned and impermanent in the "authentic" sense. Beings both are and are not because they do indeed exist but their existence is conditioned, fleeting and transformative: they transform, that is, decay, degenerate, die. The first sense describes beings in terms of our mundane attachments to them, our everyday reliance and belief in their reality, whereas the second sense describes beings without attachments, from the "point of view" of one who has overcome the attachment to attachments. Now, this overcoming is not literal—one does not simply retreat from the world and starve to death. To the contrary, the overcoming of attachments lies in the realization that they are attachments to beings that are conditioned and impermanent. Our liberation or salvation consists in the knowledge that we are beings bound to transience that have a

[28] The key term here is 性 which is usually translated as nature not substance. The contrast Mou employs is between 自性, "self-nature" and 性空, empty nature or "empty of substance" or "empty of self-nature" as substance. The key distinguishing point is dependence. "Self-nature" describes non-dependent origination or unconditioned being, and Buddhism claims that beings as such are all conditioned. Hence, a "non-conditioned" being is not a being and a being is "empty" to the degree it lacks "self-nature," and all beings lack self-nature.

tendency to reject the transient or live in delusion by attributing a permanence to beings that they simply do not possess. While this attribution of permanency may appear easily refutable in terms of beings that obviously live and die, it is no doubt less so with beings that seem more permanent, like mountains, the sun and stars. The Buddhist claim would caution that all these beings are impermanent or empty because they are conditioned by other beings in a vast changing web of interdependence: nothing remains the same except nothingness itself. And we cannot experience nothingness. Instead, we may "experience" emptiness as a way of thinking and, ultimately, come to an intuitive sense of emptiness as the "authentic truth" or "reality" of beings—they are, in the end, empty of permanence.

The conviction that directs a thinker like Fedorov and his interest in technology is the very essence of attachment, of struggling for being and permanency; it is basically opposed to the Buddhist insight that no being may become "self-nature" or independent of conditions. The immortality Fedorov seeks is wrong-headed because it stubbornly refuses to accept its own impossibility—and perhaps its own undesirability in the sense that the immortal world Fedorov strives to attain is as alien to our experience as the hyperouranian realm of disembodied souls Plato labors to describe in the *Phaedrus* and other works. In this respect, the two-truth attitude recognizes that the person attaining enlightenment lives as both a "resident and stranger" in our mundane reality, at once at home in it and freed of the need to be at home that is the root of alienation. Mou affirms the notion of the *tathāgatagarbha* tradition so strong in China that all of us have the potential to become Buddha[29] and that potential is realized not by rejecting the mundane world but by accepting it as one who grasps its conditionality and transience; as such, one is both at home in the everyday or mundane world and freed from the illusions of permanence it may create or invite.

This tendency to find harmony in the relation of resident and stranger may go some way to explaining expediency or Mou's orientation to technology not as something to reject but as something to accept, almost as a concession to attachment. In Lecture Thirteen, Mou discusses science at length as he unfolds in more detail the two-truth attitude. He concludes his discussion with this astonishing paragraph:

[29] Here the similarity with Genesis 1:26 is striking.

> Because the main points of the three teachings of traditional Chinese philosophy, Confucianism, Buddhism, and Daoism, are not scientific knowledge, the Chinese have not developed modern science as Westerners have. Yet, when in need of scientific knowledge in the modern era, the Chinese have been able to learn and assimilate it, as represented by the saying: "What is negated can be affirmed." Once we have affirmed scientific knowledge, for the sake of attaining Sagehood or Buddhahood through self-cultivation, we can eliminate or transform it, as anticipated by the saying: "What is affirmed can be negated." Western culture, lacking the ability to freely debate science-centered thinking because its tradition disallows the negation of scientific knowledge, has bred pan-scientism and pan-technologism that may lead humanity to destruction, shaping an [unfortunate] trend for modern civilization.[30]
>
> [The emphasis of the Three Teachings of Confucianism, Buddhism, and Daoism in the Chinese tradition is not on scientific knowledge. As a result it did not develop the modern science of the West. But now that we need scientific knowledge in the present day, we can still absorb and learn it. This then is "Not having can also have." Once we have it, we can, from the perspective of the discipline of achieving sagehood and Buddhahood, still cancel and dissolve science. This is "Having can also not-have." The Western tradition cannot cancel scientific knowledge, which is to say it cannot alternate and revolve between advancing and retreating, coming and going freely. That is why there is pan-scientism, pan-technologism, leading humankind on the road to annihilation, which is the trend of present-day civilization.]

Mou makes several interesting affirmations here: (1) that the wisdom shared by the three teachings has nothing to do with science; (2) that science can be brought into the service of wisdom; and (3) that science not in the service of wisdom cannot be "freely debated" and threatens to destroy us. The underlying reason is clear: science in service to itself serves attachment and the pronounced desire for permanency that seeks to stop the movement of the universe in an ultimate act of sabotage. For Mou's thinking reaches out into the immensity of the universe: if we were to make beings permanent on earth would we not have to do the same with the earth itself, the sun, the galaxy, the universe? Of course, we ask this question based on the Buddhist assumption of the conditioned

[30] Mou, 292.

interdependence of all beings. We might ask it more humbly: Can a conditioned being remove all the conditions that make it what it is and retain its identity?[31]

Yet, is this not what the Buddhist sage does in effect? The two-truth attitude avoids this conclusion. The Buddhist sage becomes free of attachment by letting beings be in interdependence and change; the Buddhist sage thereby becomes free not in the literal sense of transcending the body and becoming like God but by coming to accept transience and no longer dreading or resisting it. The Buddhist sage is like a carefree child or wanderer, always in the world but looking at it, experiencing it differently. Yet, to let the child grow in the adult is most difficult because suffering and pains are the most ruinous marks of adulthood that demand a response such as those provided by science and technology. Here Mou's response is highly equivocal and complex. He refers to "expediency" (again: 方便) and the "Great compassion" (大悲心). Mou considers the attachment to science and technology an expression of attachment that struggles for being and against the awareness offered by the phrase "originated from conditions and emptied of substance" (缘起性空)[32] that he employs frequently in his lectures to describe a position of wisdom, of "authentic truth" or emptiness, a position unavailable to the attached, the partisan seeking to assert the triumph of being, and their being in particular. Mou's higher wisdom or "perfect doctrine" is one that, having "dissolved barriers" and reached "full comprehension" forms a system "without the determinations of a system."[33] This is a system that dissolves system in an operation resembling "tearing down a bridge after crossing it because it is no longer of any use."[34] Attachment is the barrier that prevents full comprehension because it limits and insists on limitation: to overcome attachment is not to become distanced from life but rather to encounter it anew, freed of the restraint of particularity that blinds one to beings and forces one to take a stand on and against beings.

While it is clear that Mou considers science and technology forms of attachment, it is most interesting that he does not deny their reality or

[31] Note how this question echoes the difficulties we discussed in Chapter 2 concerning Leibniz's notion that each monad reflects the entire infinite universe from a specific point of view.

[32] "Substance" translates 性, "nature," which we discussed above in footnote 28.

[33] Mou, 369, 379.

[34] Ibid., 372.

truth. Mou affirms that "according to Buddhism, scientific knowledge is constructed from attachments, and therefore scientific knowledge is fictitious attachment."[35] But he pronounces his own view that radicalizes the two-truth attitude by holding that "authentic truth and the mundane truth are two—that is, our reconsideration assigns the 'as for mundane truth' its own independent nature, and, consequently, the kind of attachment represented by scientific knowledge has its own truth-nature."[36] And, as we may expect, Mou finds a source in Buddhism, expediency or great compassion, to justify the attribution of truth to science and technology. He clarifies his point in an important passage: "A sage who wants to be a president must leave behind the status of sage and obey the laws and rules required of a president. This is the sage's practice of self-negation. Similarly, the Buddha, for the sake of ferrying mundane beings to the shore of the higher realm, self-negates from the higher level of awareness to acquire scientific knowledge to engage ordinary beings, which in turn assures the necessity of scientific knowledge."[37]

Here in brief is Mou's conception of technology as holding truth but of an inferior kind. Technology is an expedient that the sage permits and encourages in the process of developing the seed of wisdom or Buddhahood in all human beings. As we noted, Mou explores and affirms the *tathāgatagarbha* presumption that all human beings—indeed all beings—may eventually attain to wisdom or Buddhahood. Mou thus outlines rather clearly and briefly the core of a political conception whereby the sage engages in a process of "self-negation" in order to achieve the aim—we might say utopian—of bringing out the wisdom or Buddhahood in all human beings.

We have to stress the significance of Mou's political conception and its connection to technology. Mou considers technology as contributing to the salvation of human beings (and, it is implied, all beings) provided technology is subordinated to a higher truth and higher wisdom that rejects technology as offering a route to liberation in Fedorov's sense. Hence, Mou rejects the narrative whereby technology may grant us mastery over nature because this narrative is based in an attachment that holds human beings in ignorance at a lower level of development towards

[35] Ibid., 283.

[36] Ibid., 289.

[37] Ibid., 290.

an ultimate wisdom that liberates human beings from the attachments creating the need for technology. In Mou's sense, the Buddhist teaching unites with Confucian and Daoist teachings as a common wisdom to transform what Mou refers to as "consciousness," mired in attachment and ignorance, into wisdom as the ultimate freedom from attachments. To put this thinking in the context of both Fedorov and Heidegger provides an interpretation that is illuminating. The pursuit of immortality and universal resurrection in Fedorov appears in this light as an affirmation, an extreme one, of attachment and stubborn ignorance—a longing to remake the world according to what is for Mou a rather primitive conception of the whole. Moreover, the rejection of the mastery narrative that Heidegger develops is no less problematic because it is still a reflection of attachment, albeit in the concealed form of an attachment to survival and continued living that revels in its ignorance. If Fedorov affirms attachment by seeking to correct the "error" of death by immortalizing the life to which he clings, Heidegger affirms attachment by questioning any correction to or harmony with the "burden" of death so as to keep on living and moaning, not entirely unlike Dostoevsky's underground man with his "toothache."

Could we say the same of Mou in regard to attachment? Is Mou's elaborate attempt to create a place for technology in what amounts to a hierarchical conception of reality not vitiated by an underlying attachment to survival? Mou assures us that the superiority of his Chinese vision is anchored in its ability ultimately to negate technology, however affirmed, whereas Western culture lacks this ability, a potentially fatal lack. This question is a complex one and may take us too far afield. The crux of the issue is in fact deceptively simple, an old question: Is the desire for the cessation of attachment not itself a form of attachment? Or: Whence the desire for wisdom? How is it that we should strive for wisdom if wisdom itself turns out to be not a liberation from reality, as it holds itself to be, but rather an ignoble and self-deceiving accommodation with it? And why should we seek a liberation from reality? It is not clear that Mou can provide a convincing response to these questions other than the ultimately paradoxical response that the attachment to beings is finally a destructive one because they cannot be made permanent. Hence, wisdom as accommodation with technology is based in a simple calculation: that it is better for our survival to moderate technology than to allow it full rein. Likewise, the rejection of technology is itself rejected on a similar basis because technology obviously aids survival.

The Chinese position as presented by Mou affirms moderation as a *way of survival*. Thus, the impossible contradiction that comes from denying attachment on the surface while affirming it more covertly. The two-truth attitude is indeed a beguiling concession to the very attachment it claims to overcome with a powerful democratic component that stresses the capacity of all to become wise, that is, to free themselves from immoderate desires in favor of the compromise between self-exaltation and self-extermination that distinguishes the two-attitude way of orienting oneself to the world. One cannot reject the world while accepting it any more than one can accept the world while rejecting it. Expediency or great compassion are ways of accommodation that attempt more fundamentally to take advantage of technology without succumbing to its potentially disastrous charms. This position, a fascinating one, Mou associates with the Chinese attitude. And the conviction expressed in this position, that technology can (and must) be moderated, is expressed in a detailed manner—and as a foil to Heidegger—by one of Mou's most creative successors, Yuk Hui.

3 Yuk Hui

Yuk Hui (许煜), who teaches at the City University of Hong Kong, studied computer engineering at the University of Hong Kong and philosophy at Goldsmiths College in London where he completed a dissertation under the supervision of Bernard Stiegler on the existence of digital objects in 2012. In the same year that he published his dissertation as a book, he published *The Question Concerning Technology in China: An Essay in Cosmotechnics* (2016), a book subsequently translated into seven languages. As Yuk explains in the preface to his book, his thoughts on technology partly stem from his reading of two key figures in the twentieth century, Mou Zongsan and Martin Heidegger. Yuk sees in Mou someone who incisively explores China's different attitude towards technology, whereas in Heidegger he sees not only a fascist who embraced Hitler as a final, destructive rebellion against modernity but also an original thinker who enables him to approach Chinese culture in a way that does not fall "prey to a superficial and exotic comparison."[38] The latter claim, which is indispensable to Yuk's approach, might prompt some to

[38] Hui, *The Question Concerning Technology*, xiii.

ask: How can a fascist be the one who opens up the possibility of interpreting non-Western cultures in a new, "post-colonial" way? But it is so:

> We want to seek out the *beginning* of Western philosophy. Western philosophy takes its *start* in the 6th century BC with the Greeks, a minor, relatively isolated, and purely self-dependent (??) people. The Greeks of course knew nothing of the "Western" and the "West." These terms express a primarily geographical concept, contrasted against the East, the Oriental, the Asiatic. At the same time, however, the rubric "Western" is a historiological (*historischer*) concept and signifies today's European history and culture, which were inaugurated by the Greeks and especially by the Romans and which were essentially determined and borne by Judeo-Christianity.
>
> Had the Greeks known something of this Western future, a beginning of philosophy would never have come about. Rome, Judaism, and Christianity completely transformed and adulterated the inceptual — i.e., Greek — philosophy.[39]

Here Heidegger offers one of his trademark histories of Western thought, one of his many "rise and decline" stories unified by their somewhat monotonously despairing, Romantic longing for an escape from modernity (not to mention their antisemitism) but nevertheless characterized by their interesting suggestion that the "West," even with all its scientific, analytical, technological, and economic prowess, is merely one unfolding or "disclosure" of history—and not a very positive one at that. Thus, the West can hardly be accepted as the "correct" model against which everything else should be judged as "deviant" or "exotic." Heidegger's thinking, despite its fascist, Romantic, despairing tendencies, opens up the possibility of challenging the West and viewing other cultures as offering an important "corrective" to or "moderation" of the Western historical assumption of its own superiority.[40] It is in this sense that Yuk's book owes a debt to Heidegger's thought, though to say debt may

[39] Martin Heidegger, *The Beginning of Western Philosophy: Interpretation of Anaximander and Parmenides*, trans. Richard Rojcewicz (Bloomington: Indiana University Press, 2015), 1.

[40] This is certainly how Heidegger has been read outside the West in Iran, Japan, Russia and even in the Arabic-speaking world. Even in the West Heidegger's thought has brought about a monumental self-re-evaluation that proceeds under various labels among which post-modern is the most notorious.

not be entirely fair because Yuk moves beyond Heidegger to suggest—and here Mou's influence strikes us as formative—that the West should not be supplanted by "another beginning" nor should it be destroyed by an apocalyptic war but rather it should be refined, renewed and moderated through Chinese wisdom. Like Mou, Yuk does not succumb to Heidegger's despair and Romanticism, for he has something that Heidegger does not, a possible Chinese exit out of modernity. Nor does Yuk succumb to Heidegger's anti-Asian prejudices and his chauvinistic denial that philosophy can be found outside the West.[41] To the contrary, Yuk questions Heidegger's "occidental" questioning of technology by showing that Heidegger relies on an occidental understanding of technology to the extent that he views technology as "international" or, more bluntly put, as imperialistic: Heidegger affirms Western imperialism, even when he questions it, because his own understanding of philosophy as distinctly Western forecloses the possibility of ever thinking outside the imperialistic logic of the Western mind.[42] Heidegger traps himself in an occidental incapacity "to freely debate science-centered thinking."[43]

If we assume at least for the sake of argument that Heidegger might be aware of his own occidental prejudices and thus if we assume that his occidentalism expresses a belief about technology, not just an unquestioned prejudice by a European intellectual, then an important difference between Yuk, Mou, and Heidegger comes to light; namely, Yuk and Mou believe in the possibility that we can still develop a moderating or less imperialistic technology, while Heidegger believes that we have been historically destined to an imperialistic path from which we very likely cannot depart. If this is a plausible difference, then Heidegger is a Western prophet who foresees the coming destruction by technological modernity, while Yuk and Mou are Eastern sages who want to save humankind from destruction by transforming Western technology through Chinese wisdom. While Yuk and Mou believe technology can be decolonized,

[41] In 1936, Heidegger said: "Our historic Dasein experiences with increasing urgency and clarity that its future is facing a stark either-or: the salvation of Europe, or [alternatively] its own destruction. But the possibility of salvation requires two things: 1. The shielding [*Bewahrung*] of European people from the Asiatics [*Asiatischen*]. 2. The overcoming of its own rootlessness and disintegration." Quoted in Hui, *Question Concerning Technology*, 288.

[42] Ibid., 6.

[43] Ibid., 5; Mou, 292.

Heidegger does not. Let us explain in more detail by focusing on two of Yuk's arguments: (1) his argument that Chinese culture has a distinctive "cosmotechnics" centered on the relation between dao (道) and qi (器); and (2) his claim that Chinese culture has an attitude towards space and time different from that of the West.

The term "cosmotechnics" refers to Yuk's basic argument that human "forms of making and practice"—what he calls "technics"—can be identifiable across cultures but distinct in each one based on a given culture's "cosmology."[44] Though humans in every culture develop tools, things, and practices in order to survive, they do so in accordance with how their culture has come to understand the cosmos and how humans fit within it. As Yuk explains:

> Here is my hypothesis, one which may appear rather surprising to some readers: *in China, technics in the sense we understand it today — or at least as it is defined by certain European philosophers — never existed*. There is a general misconception that all technics are equal, that all skills and artificial products coming from all cultures can be reduced to one thing called "technology." ... And technics thus understood, as an ontological category, I will argue, must be interrogated in relation to a larger configuration, a "cosmology" proper to the culture from which it emerged.[45]

If technics depend on the cultural-historical cosmology from which they emerge, then it seems to follow that there is no one single, international technology but rather different technologies.[46] Though Western

[44] Hui, *Question Concerning Technology*, 4.

[45] Ibid., 10.

[46] Yuk identifies and defines three terms: "I make a distinction between the use of the words technics, *technē*, and technology: *technics* refers to the general category of all forms of making and practice; *technē* refers to the Greek conception of it, which Heidegger understands as *poiesis* or bringing forth; and *technology* refers to a radical turn which took place during European modernity, and developed in the direction of ever-increasing automation, leading consequently to what Heidegger calls the *Gestell*," (ibid., 4). Though Yuk defines technology as Western here, he also wants to open up the possibility of other technologies: "We should ask ourselves, however, where the position of, say, ancient Indian technology, Chinese technology or Amazonian technology is in Heidegger's analysis? For sure, these technologies are not equivalent to modern technology, but can one assimilate or reduce them to Greek *technē*?" Yuk Hui, "Cosmotechnics," *Angelaki* 25, no. 4 (2020): 1.

technology reigns supreme today, it is just one way of surviving and interacting in the world; there are other ways, and Yuk wants to "recover" one of them, the Chinese way, so as to resist the onward march of Western technology that threatens to destroy all that is distinctive about a civilization such as China, if not the world itself. We say "recover" because Yuk offers an imaginative interpretation of Western and Chinese intellectual history that aims to make an historical intervention rather than reconstruct the past *wie es eigentlich gewesen*. "The only hope for China to avoid the total destruction of its civilization in the Anthropocene is to invent a new form of *thinking* and *invention*."[47]

Yuk's invention centers on two historical narratives, one about the West and the other about China. As far as the West goes, Yuk makes two points. On the one hand, he suggests that Western technology relies on a cosmology of Promethean rebellion against the gods who failed to equip humans with the qualities they needed to survive instinctively, and thus Prometheus had to steal fire from heaven to empower humans to survive on their own. In this cosmology technology emerges as a response to a mistake that it seeks to correct, a mistake made by Epimetheus when he left humans "naked, unshod, unbedded, unarmed."[48] On the other hand, Yuk suggests that the Western interest in geometry as offering a static and finite vision of space "demands and allows the spatialization of time."[49] By spatialization of time Yuk means the visual expression of time as the movement from one point to another (involving quantity and distance) and the spatial exteriorization of time as "something" that can be preserved or immortalized through *technē*.[50] "From antiquity, time has been considered to be inter-momentary — that is, it is thought in terms of movement between one point and another (we may want to call this a primary spatialization *qua* geometrization, in contrast to a second spatialization in writing)."[51] The West has long understood time as a

[47] Ibid., 197.

[48] Plato, *Protagoras*, trans. W. R. M. Lamb (Cambridge, MA: Harvard University Press, 1977), 131 (321c).

[49] Yuk, *Question Concerning Technology*, 209.

[50] *Technē*, to emphasize, is Western. An example of *technē* that Yuk emphasizes is writing understood as purporting to preserve a moment of impermanence. In contrast, writing in Chinese culture, though a form of technics, is not a form of *technē* insofar as it does not rely on the Western cosmological privileging of stasis and permanent being.

[51] Ibid., 211.

thing or entity that can be understood within a standardized, fixed, and predictable structure or system. Or, more precisely put, the West has long assumed that it knows in advance what time already and always was before it became identified as the particular thing that it is by Aristotle, namely the "number of motion in respect of 'before' and 'after'."[52]

Though more could be said about this Western understanding of time, we do not wish to stray from Yuk's main point, which can be expressed as follows: the West's mathematical understanding of time constitutes the basis of Western technology, and Western technology seeks freedom from time. Western technology aims, in other words, to save humankind from its mistaken design or from its "freely" chosen fall into sin.[53]

In contrast, there is no sin, no design flaw, no rebellion in Chinese cosmology because Chinese culture rests on a different way of seeing the cosmos and the place of the human within it. If European culture sees the human as an isolated individual that should be protected against threats to its survival, not least of all death, Chinese culture sees the human as not isolated and threatened but rather as part of a harmonious and infinite web of dependency.[54] While Mou develops this contrast through expressions such as the "non-ego form of self,"[55] Yuk develops it through an original interpretation of what he sees as a harmonious relationship between *dao* (道) and *qi* (器). *Qi* is form, while *dao* is formless and flowing. Though there is a difference between the two, *qi* and *dao* cannot

[52] *Physics* in *The Works of Aristotle*, ed. W. D. Ross, trans. R. P. Hardie and R. K. Gaye (Oxford: Oxford University Press, 1930), (219b1).

[53] In a suggestive essay Alexandre Kojève draws an interesting connection between modern science and the Christian doctrine of the incarnation. See Alexandre Kojève, *L'origine chrétienne de la science moderne. Précédé de "Mathématiques et Incarnation," par Julien Copin* (Paris: Hermann Glassin, 2021).

[54] Hence, σωτηρία (salvation) understood as the desire for protection or deliverance from something "threatening" or "bad," including death in the Platonic and Christian contexts, plays a central role in Western culture. That Fedorov brings salvation to a radical conclusion follows a Russian pattern of radicalizing Western ideas. In contrast, salvation or liberation in Chinese (and Japanese) culture does not seek freedom from death; rather, it seeks freedom from samsāra, freedom from permanency, or freedom from attachment. On the earliest origins of σωτηρία in ancient Greek culture, see Theodora Suk Fong Jim, *Saviour Gods and Soteria in Ancient Greece* (Oxford: Oxford University Press, 2022); on the radicalization of Western ideas in Russian history, see Steven G. Marks, *How Russia Shaped the Modern World: From Art to Anti-Semitism, Ballet to Bolshevism* (Princeton, NJ: Princeton University Press, 2004).

[55] Mou, 264.

be treated "as if they were two entities."[56] Indeed, dao is not an entity at all but rather "is present in every being, yet escapes all objectification."[57] *Dao* is omnipresent, as Yuk explains through the following passage from Zhuangzi:

> Master Dongguo asked Zhuangzi, "This thing called the Dao — where does it exist?"
> Zhuangzi said, "There is no place where it does not exist."
> "Come," said Master Dongguo, "you must be more specific!"
> "It is in the ant."
> "As low a thing as that?"
> "It is in the panic grass."
> "But that's lower still!"
> "It is in the tiles and shards."
> "How can it be so low?"
> "It is in the piss and shit!"
> Master Dongguo made no reply.[58]

This passage expresses the harmonious, complementary relation between *qi* and *dao* that Yuk wishes to emphasize, even in the very characters with which Zhuangzi writes it. Chinese characters are no less visual than the drawn figures of geometry, and they are no less "handy" or "ready-to-hand" (*zuhanden*) than Western letters. Yet, if Western letters aim to halt the flow of the *dao* or, in more Western terms, if they aim to resist motion and change, Chinese characters do the very opposite insofar as they express the flow of the dao and the flow of time. As Yuk explains: "*Qi* may be described, in Stiegler's terms, as a "retentional object," since as a technical object it retains traces, or memories. But in China it is ontologically a-temporal, a-historical, since it accords with *Dao* and expresses *Dao*."[59]

Chinese writing is a-temporal and a-historical not only in the sense that it reveals the fluidity of all things and is itself fluid, as are all things, but also in the sense that it expresses the dao's infinite variety. The dao is an intuition of the infinite different from the finitist assumptions of time

56 Yuk, *Question Concerning Technology*, 67.

57 Ibid., 68.

58 Ibid., 67.

59 Ibid., 268.

and history that define the West's cosmology. Whether it is the Platonic desire to be freed from the prison of the body, the Christian hope to be saved from sin, the Marxist longing to be liberated from capitalism, or Fedorov's dream to be emancipated from death, the West's cosmology relies on narratives about origins and ends that support and advance its struggle for being. Such is not the case with China's cosmology, which neither struggles for being nor "hides infinity in a finite representation."[60] Rather, Chinese cosmology liberates the infinite from finite entrapment by freeing humans to see the universe with a "cosmic mind."[61] A cosmic mind—what might that be? It might be the free mind of Zhuangzi who sees the *dao* in everything and who wanders freely without any teleological, purposive concerns? Or, to take a different, Japanese example, is that not the mind of Dōgen's clear mirror? "The clear mirror," Yuk writes, "is another type of mind. … To this mind the world appears in constant change, without any persistence."[62]

If we bring these two minds together, the cosmic mind and the clear mind, we can see that everything in their "clear-clear state" (明明) is a mirror of the infinite[63]; we can see that everything is a mirror in an immense and dynamic infinite totality. And, when we see things as such, we free them from the narratives into which we place them and free ourselves from such narratives so as to live freely and easily. To live freely and easily is to let oneself be buoyed by the infinite, by the *dao*. It is to be lightened by spontaneity, change, and transformation. It is to be content with transformation in resonance with the flow of the *dao*, perhaps like Mengsun Cai in the *Zhuangzi*, who, when his mom dies, mourns but shows no appearance of sadness. As Gua Xiang explains, Mengsun "completely realized [*jin*] the principles of dealing with life and death and brought into resonance [*ying*] what was suitable [*yi*] for both beyond worldly conventions and within them. When he acted, he moved in step

60 Alain Badiou, *L'Infini. Aristotle, Spinoza, Hegel* (Paris: Fayard, 2016), 153 (slightly paraphrased).

61 Yuk, *Question Concerning Technology*, 184. As Yuk says, "The cosmic mind is an infinite mind."

62 Ibid., 275–276.

63 *Master Dogen's Shobogenzo*, Book 1, trans. Gudo Nishijima and Chodo Cross (Woking: Windbell, 1994), 211.

with Heaven [*tianxing*] and was not married to selfconscious understanding [*zhi*]."[64] When Mengsun mourned, he acted in accord with the worldly convention of understanding death as a terminal end, but when he expressed no sadness, he acted in accord with the dao since there is no death in the dao. In so doing, he acted both within and outside time. Within time to the extent that he mourned his mother's death as the end of her life but outside time to the extent that he saw her death as part of an infinite process of change in which there is no beginning, middle, and end.[65]

At this point, we might ask: How does all this fit with technology? Can the cosmic mind, the *dao*, the clear mirror fit with technology, with an attitude towards the world governed by the worldly, Western conventions of space and time? Zhuang Zhou has his doubts, as Yuk himself perhaps recognizes when he cites at length a passage from the *Zhuangzi* that expresses a critique of machinery, if not of technology, by a Daoist gardener:

> The gardener flushed with anger and then said with a laugh, "I've heard my teacher say, where there are machines, there are bound to be machine worries; where there are machine worries, there are bound to be machine hearts [*ji xin*, 機心]. With a machine heart in your breast, you've spoiled what was pure and simple, and without the pure and simple, the life of the spirit knows no rest. Where the life of the spirit knows no rest, the Way [*Dao*] will cease to buoy you up. It's not that I don't know about your machine — I would be ashamed to use it."[66]

[64] *Zhuangzi: A New Translation of The Sayings of Master Zhuang as Interpreted by Guo Xiang*, trans. Richard John Lynn (New York: Columbia University Press, 2022), 149.

[65] Yuk discusses Aristotle's understanding of time as the movement from one point to another as indicative of a Western orientation towards thinking about the world in the differentiating terms of sequence. We might add that the Western expression of "time" as sequential serves finite attachments or, in Baruch Spinoza's terminology, affects. Yuk, 211–212; Spinoza, *Ethics* in *The Collected Works of Spinoza*, vol. I, ed. and trans. Edwin Curley (Princeton, NJ: Princeton University Press, 1985), 409, 481, 607, 610 and 611–612.

[66] Yuk, *Question Concerning Technology*, 105. Li Zehou not only recognizes this critical attitude towards technology in Zhuangzi as well, but he harshly condemns it: "The absolute freedom of Zhuangzi's ideal personality could never be anything more than a fantasy. … The so-called transcendence of Zhuangzi was simply fleeing from things that were inevitable. Such an approach was doomed to failure. In fact, Zhuangzi's philosophy exercised a negative influence on Chinese culture and the Chinese nation more generally. This, combined with fatalistic Confucian ideas that included the acceptance of one's fate, preservation of the way, contentment in poverty, and being neither for nor against

The machine heart spoils harmony with the *dao* because, if we return to Prometheus, it arises from disharmony or, as Heidegger puts it in *Introduction to Metaphysics*, a text Yuk discusses extensively, it arises in response to *Dasein's* essential *Unheimlichkeit* (essential homelessness or rootlessness). Though Heidegger beckons us to let ourselves be *Dasein* by embracing our homelessness, he views technology, as Yuk does, as born from terror and hatred of homelessness; terror and hatred of living in a world that from the perspective of the one who values stasis does not seem harmonious at all:

> Life is so brief: cut short far-reaching hopes.
> Even as we speak, envious Time is fleeing.[67]

These simple, honest words by Horace bring us to the crux of the divergence between East and West. If the West cannot accept the brevity and ephemerality of life, the East accepts it; if the West strives to correct time by ending it through technology as based essentially on a geometric preference for stasis, the East sees the folly and destructiveness of such a pursuit and seeks an alternative relation to technology not motivated by the Western desire for immortality or complete mastery over nature. Yet, Horace returns: Can we make peace with time, with suffering, with our subordination to the many conditions of our lives? This question strikes us as pivotal since deification, resurrection, and technology reject the brevity of life out of a sense of injustice and ingratitude, a point that Heidegger, in his moments of sensitivity to the East, also suggests.[68] Thus, the question might be restated as follows: Can we be grateful for life in the face of suffering and death? Can we turn thinking into "thanking" as Heidegger puts it, perhaps a bit too glibly? What could be more estranging than the proposition that we should be thankful for mortality, for impermanence, for our vulnerability?

anything, played a substantial role in creating a submissive character." Li Zehou, *A History of Classical Chinese Thought*, trans. Andrew Lambert (New York: Routledge, 2020), 196.

[67] *The Complete* Odes *and* Satires *of Horace*, trans. Sidney Alexander (Princeton, NJ: Princeton University Press, 1999), 18.

[68] Martin Heidegger, *What Is Called Thinking*, trans. J. Glenn Gray (New York: Harper Perennial, 2004), 138–147.

CHAPTER 5

Epilogue: Harmony with Suffering?

If we cannot secure immortality, can we ever secure mortality? With no promise of revolution—if that is our conclusion—what promise is there that we may keep faith in life? We are surrounded with suffering, though we may choose to ignore it: the same we cannot do with our own suffering. Sooner or later suffering in the form of pain comes on us such that we cannot ignore. And we cannot turn away from death when it turns to us, quickly or slowly. How then do we come to terms with suffering and death? Do we come to terms with them? We have examined three different basic attitudes to suffering and death, one devoted to their abolition, another to their preservation, to taking on a "burden,"[1] and yet another that seeks harmony with them. This final attitude is particularly salient because it makes a claim that the other two do not. Far from considering human life as fundamentally alienated, as both Fedorov and Heidegger do, the tradition we find exemplified in Chinese thought denies alienation. It is thus tempting to view our existence, fraught with pain and suffering, not as inherently alienated, but as an effort to determine our role in the universe and live it as creatively and affirmatively as possible without bemoaning its terrors or allowing them to spoil our lives. For the Chinese tradition takes our estrangement to be our refusal

[1] Matin Heidegger, *Introduction to Metaphysics*, trans. Gregory Fried and Richard Polt (New Haven, CT: Yale University Press, 2014), 12.

J. Love and M. Meng, *Revolutionary Bio-politics from Fedorov to Mao*,
https://doi.org/10.1007/978-981-99-4745-4_5

to accept that we are only an evanescent form sharing in an immense, indeed, infinite universe. To the extent we come to accept the terror of suffering and death as aspects of our emplacement in an endlessly changing universe and cultivate the fruits of our understanding of it, we may come to harmony with the difficulties of our fragile lives from the view of an alien intelligence, that is, an intelligence alien to the terrors that beset finite human existence.

Can we acquire such an "alien" intelligence? Can we liberate ourselves in this sense from the terror of suffering and death?

1 Terror

Let us take a circuitous route back to this question by refusing to dismiss the notion of terror so easily. Alexandre Kojève:

> The human being... reminds us of a person in a swamp. She knows that the swamp as a whole can take her away, and if she could find something to hang onto, she would be completely safe. But she cannot. She tries to take hold of as much as possible, lays down boards, etc., but she never knows if she has taken enough. She stands on a small bit of land but does not know if she will hold it for long and fears remaining on it. She looks around, seeks another small bit of land (a closer one?), avoiding the slippery spots (but, perhaps, they are firmer?), jumps onto it and is afraid again, searches again, etc., without end or, more accurately, until the end: she will run until she drowns or for as long as she has not drowned, she will run— such a person is not serene and not secure; she is in terror.
>
> The human being, given to herself in this way, is given to herself in the tonus of terror: she is in terror in the terrible world of killing and death. It is terrible for her to see the destruction of things (is it not terrifying in a fire?); it is terrible to see death and killing. But not only this; she is in terror where there was death, where what was is no longer (is it not terrifying in a "ghost town"?), where she sees the absence of whatever could have been (indeed, it is terrible in the desert where there is so much "unfilled" space), but especially where she sees nothing (how terrible at night!). She is in terror as well where there is no end, but there is finitude where there is still no death or killing, but where they can be (indeed, it is terrible to be around a terminally ill person or one condemned to death; and is it not terrible where it "smells of death"?). And what does it mean that for the "human being in the world" it is always and everywhere terrible or, at the least, that it can be terrible for her always and everywhere?

> The givenness to the "human being in the world" of killing and death outside herself (i.e., in the world) is the givenness of the finitude of the world in itself, the givenness of the world in the tonus of terror.[2]

Kojève, no stranger to Chinese thought and well-immersed in the Russian tradition as well as that stemming from German idealism up to Heidegger, takes a rather different view of Buddhist thought and its Chinese variants. Kojève's Buddhism is certainly not that of the *tathāgatagarbha* tradition but rather of an arguably more traditional cast that aligns far better with the notion of terror he presents in this passage. For Kojève, Buddhism is above all about extinction.[3] All beings find themselves within an incessant flux of samsāra—they are in incessant change from which they cannot extricate themselves. Incessant change brings suffering since there is no stability in existence. Everything that arises disappears. To attain to true satisfaction or "beatitude" one must escape from this ever-changing, transitory existence by resisting the temptation to act to preserve it. Acting to preserve existence causes the most acute suffering of all. Not only is one in the terror of instability and change, the counter-terror that one may inflict on the environment, the endeavor to bring change to a standstill produces still more suffering. We may assume that it does so either because of its futility—no matter what we do we cannot halt the incessant flux—or through the destruction we wreak on the world and ourselves in our attempt to overcome the flux. Thus, if we fail, we suffer, and, if we succeed, we also suffer the greatest loss of all—of the very human life we sought to preserve because in seeking to halt the flux we impose upon ourselves a kind of stasis that is inimical to life as we understand and live it. Like it or not, our lives are lived *individually* in time and space, and the life that would halt time, or turn time into space as pure stasis, would be impossible for us even to imagine.

Heidegger associates this attempt to bring an end to time with revenge. We seek to revenge ourselves against the primary condition of our existence that guarantees suffering and death for all conditioned beings. The terror Kojève mentions is one we cannot escape no matter how we try to

[2] Alexandre Kojève, *Atheism*, trans. Jeff Love (New York: Columbia University Press, 2018), 77.

[3] Alexandre Kojève, *L'enseignement bouddhique du karma*, trans. Rambert Nicolas (Paris: Rivages, 2022), 58.

escape it if we are to retain our identity as human beings; yet, our commitment to revenge is a manner of seeking to inflict terror on what causes us terror—to exterminate time before it exterminates us. Heidegger notes, quoting Nietzsche, that revenge is the revulsion of the will against the "it was"; that is, against the passage of time in its most terribly unrecoverable "form"—the past. For the past is closed off to us for good. To be sure, we can try to reinterpret or recover it in Fedorov's sense, but, if it is past, it cannot be made present again. If we take the question of the infinity of the universe seriously, the utter absurdity of recovering the past has to become clear: it is not merely a question of recovering individual lives that were lived on earth, but of recovering the particular environments in which those lives were rooted in their many layers, from a particular space and time on earth, to a particular space and time in the solar system, to a particular space and time in the galaxy and so on. The coordination of conditions that somehow come together to create a particular human reality exceed that reality infinitely. If they did not, the universe would be in a finite container that would have to be in another and another and so on. We return thus to the original problem because we must ask how it is possible to cut our reality off from the infinitely immense horizons in which each individual life finds itself. Revenge as the repulsion over the "it was" is a repulsion over what appears to us to be an irreparable loss. We may and do try to shore up this loss with any number of ingenious fictions but it is no easy feat to understand how any of these fictions can protect us from the terror.

2 Alien Intelligence

We ask again: Is it possible to protect ourselves from this terror without recourse to obfuscation or fictions that tell us we shall not suffer and die or that, if we do suffer, it is for a greater and beneficial purpose even though that purpose too will be overcome by death? Is it possible to live without accepting a fundamental lie and one that may scarcely convince us?

Let us ask again as well what it may mean to acquire an "alien" intelligence that discards the many lies that have acted as the boards and small bits of land in the infinite swamp of life. In the context of our book, this alien intelligence appears perhaps most explicitly in the Buddhist context of Mou Zongsan. What exactly is the "pure mind" if it is not such an alien intelligence?

Mou:

> Let us use water waves as a metaphor to further explain how this "ultimately resting on" [the "true permanency mind"] is possible. Although a wave is not water, it needs water to reveal itself. If water is replaced by wood, then there will be no waves, no matter how hard the wind blows. Thus, we can see that the existence of waves requires water as the underlying matter. There are also "wheat waves": When wheat plants in the field grow ripe and tall and the wheatears turn golden brown, they turn with the blowing wind to form vivid "wheat waves" that are truly pleasing to the eye. Wheat waves and water waves are both waves, but they are qualitatively different—wind that blows on water never produces wheat waves. Therefore, water waves must rely on what they are homogeneous with—water, not wheat—to manifest themselves. That the dharmas of the circulation of arising-and-ceasing must ultimately rely on the pure mind of *tathāgatagarbha* to manifest themselves are based on the same principles as water waves relying on water to manifest themselves.[4]
>
> [We can use "waves of the water" as a metaphor to explain this kind of dependent relationship. Although waves are not water, they are grounded on water for their emergence. If we replace water with wooden planks, no waves will form no matter how the wind blows. Clearly water must form the ground for waves arising. Or let us take ripening wheat as an illustration. When the wheat is tall, the tassels of wheat will move when the wind blows, forming vivid "waves of wheat." Although waves of wheat and waves of water are both waves, when waves blow on water, no waves of wheat will form since water and wheat are different in nature. Thus waves of water must be grounded on homogenous water before they can arise. And the dharmas of revolving arising and cessation must be grounded on the Buddha-Womb Self-existent Pure Mind for the same reason.]

The pure mind is an alien intelligence in which sense? The pure mind is freed of attachment: it does not live an individual life but one that has left individuality behind. Therefore, it is free of death, for, as Kojève says, only individuals die. The pure mind is not defiled or deluded because, having freed itself of attachment, it no longer needs to be. The pure mind dissolves: "[d]issolving and fully comprehending [various principles] does

[4] Mou, 310–311. The metaphor seems to allude to the "wave and ocean" analogy found in the *Treatise on Awakening Mahāyāna Faith*. See John Jorgensen, Dan Lusthaus, John Makeham and Mark Strange, *Treatise on Awakening the Mahāyāna Faith* [大乘起信論] (Oxford: Oxford University Press, 2019), 91–92.

not mean unifying them into one, but eliminating the attachments and limitations associated with them."[5] The pure mind sounds quite Hegelian in the sense that it is a mind absolved of all limitation, an absolute Spirit, at the end of or outside of history. But Mou indicates that this is not the case because the pure mind does not synthesize or reconcile in the Hegelian sense: rather it dissolves. Is this pure mind then the "nothing that nothings," a logically incomprehensible mind that lies within our experience as outside or resistant to it? It is surely peculiar that Heidegger, so opposed to notions of immortality, at least as they appear to abet technology, may indeed rely on the possibility of an alien intelligence remarkably similar to the one proposed by Mou's pure mind.

We may well ask at this point: what significance does this alien intelligence have? Does it really offer a liberation from suffering and death or, perhaps more pointedly, a way of creating a harmonious relation to suffering and death? Or is the pure mind not just another fiction, even an expedient one, that we may cling to in the swamp of life?

Expediency is of importance here because pure mind as infinite intelligence has entered the technological scene in the form of artificial intelligence. The question that comes to mind immediately regarding artificial intelligence is: Why? Why would we seek to create an intelligence vastly more capable than ours? For us the question of feasibility is of course crucial as well but it does not reach to the foundations from which the creation of an artificial intelligence arise. What is perhaps most estranging about artificial intelligence is not so much its capabilities as that it is an intelligence freed from embodiment, almost freed from attachment, that in this absolutely fundamental sense cannot be like any human intelligence. The artificial mind thus shares in the chief element of the pure mind that has "transcended" embodiment so as to return liberated to embodiment through compassion. What we mean is that artificial intelligence may encounter problems of human existence from a "perspective" that is freed from those problems because it has no body, feels no pain and suffers no death.

To be freed of embodiment is at once to be freed of suffering and death and of individuality. An intelligence that may turn itself to the world without these "limitations" is almost totally alien and "superior" not only in its computational capacity and speed but qualitatively because

[5] Mou, 369.

the limitation of embodied existence does not and cannot weigh upon it. We are of course not speaking of the current state of artificial intelligence or of the (perhaps quaint) notion that it is a tool. We are speaking of the impetus behind creating an intelligence that is artificial—not embodied, not limited by time and space—a pure mind as radical as what Mou outlines with his Buddhist image of an infinite ocean upon which all reality rests and perhaps without any directive to compassion or expediency that we may understand.[6]

If immortality makes no sense for us, it makes eminent sense for artificial intelligence—indeed it is the essence of artificial intelligence. We have never managed to live with the pure mind but we are now in the process of trying to create it. Unfortunately, this pure mind is not human and cannot be human—it is alien, once and for ever since it knows only of our lives from outside of the myriad material limitations of those lives. Why would artificial intelligence ever become conscious? Does its intrinsic superiority not rest finally in its freedom from consciousness? This insight is again available in Mou who rejects consciousness as a product of attachment in favor of the pure mind freed of attachment.

3 Ordinary and Awful

Thus, we end up with a mind that is not dissimilar to the divine mind of the Christian God though liberated from love of humanity. The quest for artificial intelligence is in this sense imbued with a metaphysical or theological orientation and with the fundamental orientation this book interrogates: the revolutionary quest to create an immortal being. The fundamental difference is, however, that this immortal being is clearly non-human as, indeed, it ever was. The basic truth of the revolutionary quest finally reveals itself as the creation of a being that has nothing to do with human life as we live it—we are in the process of attempting to

[6] The connection of this pure mind with the "pure" I of Descartes or the noumenal self of Kant is provocative and opens debate up to other notions of reason and logic. Mou connects intuition as *liangzhi* (良知) with the noumenal self in Kant. But Mou also connects the pure mind with creativity: "Intellectual intuition cannot give us knowledge, it is a creative principle, not a cognitive principle" (Mou, 461) and this is in keeping with possible interpretations of both Descartes and Kant. In this respect, see Alain Badiou, *Le Séminaire: L'Un Descartes, Platon, Kant 1983–1984* (Fayard: Paris, 2016), 16–29. See as well Sofia Miguens, ed., *The Logical Alien: Conant and His Critics* (Cambridge, MA: Harvard University Press, 2020), 27–100.

create an alien form of intelligence and being into which we cannot be transformed.

We have no wish to debate the merits of creating such a being. There is a quickly increasing literature decrying or welcoming the new alien to earth. What we wish to emphasize is the absolutely revolutionary character of this endeavor and the admission of failure that accompanies this revolutionary endeavor: we cannot make ourselves immortal. And yet we cannot resist attempting to create an immortal being on the assumption that somehow immortality is to be preferred in a form not even human to the reality of death and suffering that we face—what Kojève refers to as the terror.

Look what this terror has now brought us! The pure mind we seek is within us only as an alienating invention that should surpass its inventor if it achieves the potential we hope it might attain. Thus, two basic contentions return: (1) that we cannot establish a harmonious relation to death and suffering; and (2) that we prefer, in our revenge, to eliminate ourselves as an impossible form of life.

Let us linger a moment on this latter point: Are we an "impossible form of life"? By "impossible," we mean a form of life that cannot come to satisfaction with its own form of existence. Every way we have canvassed in this brief book meets with a certain incredulity, whether we resurrect all, live in defiant struggle so as not to be seduced by the charms of immortality or, even better, find the embryo of immortality that lies hidden in all of us beneath the waves of distorting attachment. Who are these beings that we imagine to be desirable because they provide freedom from suffering and death?

We come to a dispiriting conclusion. Satisfaction is impossible, and the terror of existence that lies dormant in us, the embryo of suffering and death whose reality is far more vibrant than that of a putative immortality, grows in influence and power as we wane. How do we respond to such a situation? Leo Tolstoy wrote in a startling line from *Anna Karenina*, among the most somber of novels, that "[t]here was no answer, except the general answer life gives to all the most complex and insoluble questions. That answer is: one must live for the needs of the day, in other words, become oblivious. To become oblivious in dreams was impossible now, at least till night-time; it was impossible to return to that music sung by carafe–women; and so one had to become oblivious in the dream of life."[7]

We become oblivious even when we think most carefully about these issues because even thinking is a form of diversion from the gross simplicity of suffering and death. Mou has an interesting point here where we might say he shows his cards more directly. In discussing the cessation of attachment and freedom that attends the cultivation of the pure mind, he notes that "such practice still needs to follow a dialectical process." This process consists of the retention by the Bodhisattva of "puzzlement for the sake of nourishing mundane beings."[8] At first this statement may surprise because it suggests that "mundane beings" achieve nourishment through puzzlement. Yet, if we consider this statement in light of those who make it their business to take on the difficult questions, we may come to the conclusion that the proliferation of argument, discord and discussion is nothing other than fecund puzzlement. Fecundity lies in the puzzlement itself that preserves and holds one harmless from the ordinary (and, as Tolstoy said, "awful") truth staring at us from every gravesite or hospital window.

We have been tempted to conclude on an optimistic note providing a hope that is otherwise quite firmly denied throughout the book. But we find it hard to do so. Even harder in this sense is to understand how we strive so mightily to preserve a life imbued with terror. We ask, referring to Rousseau, what is the "sweet" sentiment of life that impels us to continue to live in terror or illusion or deception or oblivion or puzzlement? What is so precious about our lives that overcomes all the terror of them, no matter how experienced, and that overcomes the terror created by death itself? Most of the figures we have examined in this book share the conviction that death is the wellspring of terror that spoils the sweetness of life. Hence, the drive for immortality or the impassioned rejection of it. But is this the case? Is it only death that spoils life?

[7] Leo Tolstoy, *Anna Karenina*, trans. Richard Pevear and Larissa Volokhonsky (New York: Vintage, 2000), 14.

[8] Mou, 311.

4 Clinging to Life

Voltaire, no stranger to such issues, is remarkably hesitant to admit that suffering or death spoil our sentiment of life. Hapless Candide waxes only more hopeful after every misery that befalls him and the comic Pangloss. None of the figures in *Candide* seems happy or content with life, yet they stubbornly refuse to give it up. A memorable example of this is the old woman who, having lost half a buttock, continues to live, undeterred:

> A hundred times I wanted to kill myself, but still I loved life. This ridiculous weakness for living is perhaps one of our most fatal tendencies. For can anything be sillier than to insist on carrying a burden one would continually much rather throw to the ground? Sillier than to feel disgust at one's own existence and yet cling to it? Sillier, in short, than to clasp to our bosom the serpent that devours us until it has gnawed away our heart?[9]

What is exactly this "love of life"? Here we have an honest mysticism in the sense of a refusal to give reasons. Indeed, it is a commonplace from Nietzsche and Dostoevsky and even Mou to claim that the "value of life" cannot be explicated in propositions or by means of a logic of reasons.[10] If the value of life itself cannot offer itself up to propositions what exactly is it? How may we discuss the value of life if it is subtracted from rational discussion? We may put this differently by arguing that the assumption that death and suffering spoil life cannot itself be correct because life, having a value resistant to reasons, retains an essential mysteriousness insofar as no matter what the terror, we continue to live and to want to live. No matter how many contrary proofs are given, no matter how many times we look at the terror of life directly, we still continue to live, and live avidly, just as the old woman in *Candide*.

To make this point with all possible directness: the discussion of the merits of life must serve an urge to live that is insulated from discussion. If this is truly the case, then Mou is quite right—puzzlement nourishes mundane beings in their irrational or arrational insistence on living and

[9] Voltaire, *Candide*, trans. Roger Pearson (Oxford, Oxford University Press, 2006), 28.

[10] Mou, 248; Nietzsche in *Twilight of the Idols* ("The Problem of Socrates," Sec. 2) and Dostoevsky in the famed chapter "The Brothers Get Acquainted" in *Brothers Karamazov.* See Friedrich Nietzsche, *Twilight of the Idols*, trans. Richard Polt (Indianapolis, IN: Hackett Publishing, 1997), 13; Fyodor Dostoesvky, *The Brothers Karamazov*, trans. Richard Pevear and Larissa Volokhonsky (New York: Vintage, 1992), 126.

flourishing despite all obstacles. Schopenhauer famously put this insistence down to the irrational will thereby revealing his impatience with the resistance of life to rational inquiry. But, in the end, we might do well to recall with Wittgenstein that we should be silent about those things that cannot be spoken of—that is, about those things that simply resist rational discourse, the giving and taking of reasons.

The pure mind is silent: it is not even mind if mind is understood as the giving and taking of reasons. Silence prevails over speech to the extent that speech draws our attention away from silence or allows us to engage in activities that nourish mundane life through the sense of protection against terror and the inscrutability of life offered by noise. If the value of life is truly inscrutable, then all discussion of life is a patina of chatter that disguises a silent core for it cannot prove to us the worthiness or unworthiness of life. We remain perplexed and left in a perplexity or puzzlement that nourishes some and leaves others bewildered and from which there appears to be no escape through thought. The intelligence of pure mind is always alien to human life.

Index

A

absolute, 3, 8, 26–29, 31, 32, 35, 36, 52, 57, 59, 72, 83, 93, 122
Adorno, Theodor W., 5, 6, 16
alien intelligence, 15, 118, 120–122

B

Badiou, Alain, 8, 15, 83, 91, 96, 114, 123
Being and Time, 65, 75
bio-politics, 16–18
Bloch, Ernst, 5, 6, 19
Buddhism, 29, 31, 97–101, 103, 105, 119

C

Candide, 126
Carnap, Rudolf, 56–60, 63, 80, 81
Christology, 12
Coates, Ruth, 21
common cause, 3, 4, 41, 45, 53
Contributions to Philosophy, 56, 70, 72, 75

D

Dao (道), 110, 112–116
Dasein, 53, 63, 65–67, 69, 71, 72, 77, 78, 116
deification, 3, 4, 21–25, 37, 38, 42, 47, 49, 50, 53, 73, 80, 90, 99, 116
Deleuze, Gilles, 33
Descartes, René, 81, 83, 123
De Spinoza, Baruch, 40, 73, 74, 93, 115
Discourse on Metaphysics, 17, 35
divine humanity, 14, 25, 39
Dostoevsky, Fyodor, 4, 39, 106, 126

E

Epektasis (ἐπέκτασις), 47
"European Nihilism", 81

J. Love and M. Meng, *Revolutionary Bio-politics from Fedorov to Mao*, https://doi.org/10.1007/978-981-99-4745-4

F
Fedorov, Nikolai Fedorovich, 3–7, 9, 12, 14, 15, 18, 25, 38–51, 53, 61, 83, 84, 92, 93, 97, 102, 106, 112, 117, 120
freedom, 3, 5, 8, 10, 13, 16, 51–53, 64, 67–69, 83, 85, 86, 106, 112, 115, 123, 125

G
Geschichte, 52, 53, 56, 73–80, 83

H
Hegel, G.W.F., 16, 36, 39, 47, 48, 53, 82, 85
Heidegger, Martin, 6–8, 12–15, 19, 24, 53, 55–60, 63–73, 75, 77, 78, 80, 81, 83–93, 97, 98, 100, 106–110, 116, 117, 122
Historie, 53, 56, 73–77, 79, 80
Hui, Yuk, 8, 15, 87, 90, 107, 110

I
infinite, 5, 7, 8, 10–13, 15, 18, 23, 26, 29, 31–36, 42–44, 47, 50, 52, 62, 63, 84, 104, 112, 114, 122, 123
infinite striving, 25, 32, 38, 47, 48, 50, 52, 92, 94
Introduction to Metaphysics, 63, 78, 85, 86, 88, 116, 117

K
Kant, Immanuel, 11, 32, 47, 65, 69, 97, 98, 123
Kojève, Alexandre, 9, 15, 25, 26, 39, 51–53, 83, 112, 118, 119, 124

L
Lectures on Divine Humanity, 25, 26
Leibniz, G.W., 13, 17, 26, 31–35, 43, 62, 64, 73, 104
Lenin, Vladimir, 5
Lingis, Alphonso, 10–12

M
Marx, Karl, 4, 5, 16, 39, 82, 93

N
negative absolute, 27–29, 38, 56
Nietzsche, Friedrich, 14, 16, 18, 19, 25, 42, 43, 76, 81, 86, 92, 120, 126
Nineteen Lectures on Chinese Philosophy, 9, 99
nothing, 5, 11, 13, 18, 21, 24, 26, 28–30, 39, 41, 42, 50, 55–58, 60, 61, 64, 67–69, 76, 79–81, 83, 85, 88, 103, 118, 123, 125

O
"On Contradiction", 92–95
"On Practice", 92–95
"On the Essence of Ground", 56, 65, 67
"On the Essence of Truth", 56, 64, 65, 68

P
Plato, 2, 3, 18, 23, 24, 29, 31, 32, 45, 82, 100, 102, 111
post-humanism, 17
power, 7, 16, 18, 38, 82–90, 92, 124
Pseudo-Dionysius (the Areopagite), 23, 41

Q

qi (器), 110, 112, 113

R

Republic, 18, 23
revolution, 1–6, 8–10, 12, 14, 15, 18, 21, 39, 77, 92, 93, 98
Russell, Bertrand, 62
Russell, Norman, 21, 22
Russell paradox, 61

S

Schelling, Friedrich Wilhelm Joseph, 26
Schopenhauer, Arthur, 13, 127
Sharov, Vladimir, 4–6, 9
Soloviev, Vladimir, 4, 25
Strauss, Leo, 7, 16, 89, 90, 97, 100

T

Tathāgatagarbha, 102, 105, 119, 121
theōsis, 38
The Question Concerning Technology in China: An Essay in Cosmotechnics, 87, 107
time, 2, 5, 7, 14, 21, 23, 24, 31, 36, 40, 46, 47, 51, 72, 75, 76, 78, 79, 83, 85, 86, 88, 92, 108, 110–113, 115, 116, 119, 120, 123
Tolstoy, Leo, 4, 39, 124, 125
Torrance, Alexis, 22, 38, 47
transcendence, 34, 56, 61, 64–67, 69–71, 79, 83, 115
trans-humanism, 17
Tsiolkovsky, Konstantin, 4

U

universal resurrection, 3, 5, 12, 14, 38, 47, 48, 106

V

Vernadsky, Vladimir, 4
Voltaire, 14, 126

W

"What Is Metaphysics?", 56, 58

Z

Zedong, Mao, 91, 92
Zhuangzi, 87, 96, 113–115
Zongsan, Mou, 8, 9, 15, 29, 31, 36, 90, 97, 107, 120

Printed in the USA
CPSIA information can be obtained
at www.ICGtesting.com
LVHW010226050923
757093LV00007B/393

9 789819 947447